To Minesh

OUT OF

THE ASHES

with my best wishes
of success

W macetti

OUT OF
THE ASHES

Tools For Recovering
Corporate Health

Jean Frédéric Mognetti

John Wiley & Sons, Ltd

Other Wiley Editorial Offices

John Wiley & Sons Inc., 111 River Street, Hoboken, NJ 07030, USA

Jossey-Bass, 989 Market Street, San Francisco, CA 94103-1741, USA

Wiley-VCH Verlag GmbH, Boschstr. 12, D-69469 Weinheim, Germany

John Wiley & Sons Australia Ltd, 42 McDougall Street, Milton, Queensland 4064, Australia

John Wiley & Sons (Asia) Pte Ltd, 2 Clementi Loop #02-01, Jin Xing Distripark, Singapore
129809

John Wiley & Sons Canada Ltd, 6045 Freemont Blvd, Mississauga, ONT, L5R 4J3, Canada

Wiley also publishes its books in a variety of electronic formats. Some content that appears
in print may not be available in electronic books.

Library of Congress Cataloging-in-Publication Data

Mognetti, Jean-Frédéric.
   Out of the ashes : tools for recovering corporate health / by Jean Frédéric Mognetti.
   p. cm.
   Includes bibliographical references and index.
   ISBN 978-0-470-51868-7 (cloth : alk. paper)
   1. Corporate turnarounds.   2. Organizational change.   3. Airlines – Management.
I. Title.
   HD58.8.M596   2008
   658.4′06 – dc22

                   2007046842

British Library Cataloguing in Publication Data

A catalogue record for this book is available from the British Library

ISBN 978-0-470-51868-7 (HB)

Typeset in 11.5/15pt Bembo by SNP Best-set Typesetter Ltd., Hong Kong
Printed and bound in Great Britain by TJ International Ltd, Padstow, Cornwall, UK

CONTENTS

FOREWORD

Richard Lapthorne,
Chairman, Cable and Wireless plc

There is nothing self-evident about fostering a corporate dynamic that boosts the value of shareholders' investment, nor in reserving capital requirements for the next phase of development. Under the influence of private equity firms, a productivity dimension has been added to good strategic management and execution by asking a straightforward question: at what cost? Successful CEOs now have not only to address the affordability of any relevant expenditure, but also how to optimize its cost of funds employed for its settlement. This strong desire to become more self-sufficient requires them to be even more creative and systematic in designing the sources of funding. Simultaneously, it also tests the CEO's own active commitment to boosting corporate value and making appropriate trade-offs between short-term cash performance and long-term value. An expenditure that will substantially cut your own bonus is an unambiguous acid test to confirm that all interests are well aligned on a common target.

For too many years CEOs have enjoyed limited pressure for high return on equity and too much attention has been paid to P&L, which is good but insufficient. A balance sheet management approach at board as well as executive committee level is far more relevant in terms of both performance and ambitions assessment. It permanently addresses the question: will we have the resources to reach our future earnings ambitions? However, in the case of a turnaround, do such sound principles remain relevant, or must they be temporarily put in brackets until the company is back on a healthier growth track?

Out of the Ashes is based on Jean-Frédéric Mognetti's hands-on experience at SN Brussels Airlines. He proposes an original and lively answer to devising healthy management principles. His book shows how building a positive management development context from the bottom up immediately following a bankruptcy produces the conditions for future balance sheet management, giving the same meaning to performance assessment from its lowest level of observation, a flight, right up to the executive committee.

Management tools and solutions are suboptimized unless they are assimilated and shared by the whole organization. This is a constant quest throughout this book, which allows you to dive into the day-to-day reality of transforming the management reflexes and behaviours of this company. In this saga, Mognetti found a way to help the whole company with a special focus on the business unit manager becoming steeped in performance assessment, assimilating everything from variable costs coverage to asset renewal contribution (aircraft fleet) with an accessible, transparent, real-time performance indicator or corporate pulse, named RoDoC, standing for return on direct operating costs. The simple three-milestone system associated with RoDoC is surprising in its simplicity. Based on its results, it quickly demonstrates an original management sleight of hand to instil relevance and transparency, which operate as the cement that consolidates the confidence and trust-building system introduced by the Executive Chairman and the CEO. The whole company has daily access to the current stage of the recovery. Everyone knows where they stand in relation to the quality of current performance, but also how the company is strengthening itself to embrace its future ambitions.

Finding the appropriate performance measurement tool is the first step in Mognetti's proposal; defining its relevant use according to the stage of evolution of the company is another question addressed in this management story. For instance, Mognetti was able to make the whole team conscious of one key question: do you generate an operating contribution when performing your *raison d'être*, flying from A to B? Without hype or any big announcement, a large-scale dialogue system, called the Route Business Strategy Meeting (RBS), recreated a sense of profitability improvement among all performance stakeholders. The details of its design show the discipline needed to achieve the required transformation in management reflexes and attitudes. Moreover, the dynamics of transformation do not stop at the assessment of past performance. They also create the conditions to move ahead, cast the company into its future, and deal with uncertainty by minimizing the risk. Thanks to an original dialogue-based forecasting approach, Mognetti avoids the trap of a simple extended budget, emphasizing the question of the appropriate timeframe for this dialogue about the future.

Out of the Ashes follows a pleasant structure. It alternates a management journey through the day-to-day reality of a live example with commentaries by chairmen and CEOs to broaden ideas generated by his business transformation proposal. This enables Mognetti to help the reader find their own best way to operate in the best interests of their customers, staff and shareholders.

The last question in the book confirms that balance sheet management starts from the very first day of any attempt at recovery. What, Mognetti asks, is the difference between a turnaround and a remission? The unambiguous answer: the ability to keep control of one's destiny through having the capacity to renew one's assets. If one cannot do this, suitable exit conditions must be anticipated well in advance, again to protect the value of the company. This lucidity is at the heart of a pragmatic management attitude.

INTRODUCTION

On 22 September 2005, Rob Kuijpers, Executive Chairman of SN Brussels Airlines, was the guest speaker at a conference organized by accountants Ernst and Young in Brussels. It was the day after he had sent his dismissal letter to the President of the board, but understandably in this context he did not make any allusion to his dismissal. Nor did the audience. The conference was heading for its end when someone asked for the microphone, stood up and said, 'Mr Kuijpers, I am an average Belgian citizen. Thanks to your achievement at the head of SN, we know that our taxes are no longer being used to subsidize a chronically unprofitable airline company, and I wanted to say you thank you on behalf of my compatriots.' This triggered a standing ovation.

In order to deconstruct and examine the management dynamics and skills behind this spontaneous moment of corporate truth, we need to be back in Fall 2001. That was the time when SABENA, the Belgian national carrier, went bankrupt, along with its mother company Swissair. This seemed like a veritable earthquake in the airline industry. The pride of both countries was deeply wounded. Airlines are like flags, so a 'phoenix plan' was immediately engineered to relaunch the national carriers out of the ashes.

Nevertheless, in the Belgian context national pride had already cost a great deal. SABENA – the second oldest European airline company – had made a profit only twice in its more than 70 years of existence: first in the year of the Universal Exposition in Brussels in 1958; second in 1998 when Swissair changed some accounting principles after the merger of the two airlines. So in Belgium in mid-2001 the paramount feeling was 'never again'. Under the leadership of Viscount Etienne Davignon, one of the most prominent and influential businesspeople in his home country and beyond, a national solution was quickly established involving more than 40 private and institutional investors. This 'start-through' or relaunch with the strong flavour of a turnaround sharply downsized the scale of the airline's operations, leading to reductions of 80 per cent in staff numbers, 50 per cent in aircraft and 33 per cent in passenger numbers. This was not at all comparable with the magnitude of Swiss ambitions at the same time.

Peter Davies, Chief Operating Officer of logistics company DHL, was foreseen as CEO of this new company, while his former CEO at DHL, Rob Kuijpers, was appointed Executive Chairman. Peter accepted the offer in May 2002 and called me for support in some early relaunch tasks. Beginning as quick initial assistance, my contribution turned into my becoming shadow Restructuring CataPyst for more than 40 months, incidentally putting on hold my involvement in academic work as of mid-2004.

In May 2002, the results of my past consulting and research collaborations with DHL and other leading service companies operating in a business-to-business environment, such as Sodexho, had just been published by John Wiley in a book called *Organic*

Growth. But my practical, experience-based recommendations and skills in the area of growth were not at all appropriate in this newly reborn company, staffed by 1800 traumatized people. A radically different toolkit needed to be developed to immediately overcome two inherent pitfalls of the current situation:

• Staying stuck in the nightmare of bankruptcy.
• Replicating the former tailspin pattern, not breaking free from the past.

This represented my daily playing field: developing new management skills, techniques and attitudes to put into action the hope that the relaunch of this company had generated among its employees, its clients and its shareholders.

Even while I was involved in this 'hands-on' mission, I kept my academic genes alive and incorporated this form of practical research into a second behind-the-scenes story about recreating a context of healthy growth after such a spectacular business accident. Creating a context means what we do so that our wishes and ambitions have a real chance of becoming tangible reality. Context building is about tenacity and discipline in simple things. The following comment from Anne Mulcahy, CEO of Xerox, gives a flavour of what this task was about:

> I was doing a customer breakfast in Dallas. We have invited a set of business leaders there. One was a plainspoken, self-made, streetwise guy. He came up to me and gave me this advice, and I have wound up using it constantly. 'When everything gets really complicated and you feel overwhelmed,' he told me, 'think about it this way: You gotta to do three things. First, get the cow out of the ditch. Second, find how the cow got into the ditch. Third, make sure you do whatever it takes so the cow doesn't go into the ditch again.' Now, every time I talk about the turnaround at Xerox, I start with the cow in the ditch.[1]

[1] Anne Mulcahy (2005) 'The best advice I ever got', *Fortune*, March 21.

SN Brussels Airlines was my equivalent of Anne Mulcahy's 'cow in the ditch', and everything had to be done so that it didn't return to the ditch. This book is about the efforts of an entire organization to do just that. As in *Organic Growth*, my goal is to take the reader behind the scenes in order to share the day-to-day details of what was imagined, formulated, implemented and adjusted, with the result that this company posted a profit in only its second year of operation, and was even able to acquire Virgin Express in 2005. The question is what management lessons we can draw from this recovered health and apply beyond the airline sector. How, with 12 times less equity investment than Swiss International Airlines, has this company managed to stay independent, while its former sister company was acquired by Lufthansa for peanuts after having burnt through nearly €1 billion in its first year of operation.

Business comebacks like this are not the result of a spectacular strategic insight. SN Brussels Airlines has a clear description of its strategy, 'connecting Brussels', but the management flavour of a 'start-through' is a combination of focusing on people and creating tailor-made management tools. These tools are not sophisticated, but they do fit perfectly with the context of this specific business recovery. The CEO's role in creating two elements

- a two-pronged trust- and confidence-building mechanism, downwards to the company and its management, upwards to the board and the shareholders
- an appropriate context, to help people operate to the best of their abilities

offers some specific management lessons that are transferable to many business contexts.

Some basic management tools that had become hidden under layers of accumulated routine were reignited to rejuvenate their true management meanings. Once this dynamic had begun, the cultural transformation process revisited an area that had been taken for granted, the fundamentals of performance assessment. A disruptive corporate performance unit, RoDoC (return on direct operating costs), became the new pulse of this company. The company's management learned how to speak a different language and developed a systematic dialogue to increase the relevance of its diagnosis and its speed of reaction. Once the new toolkit was fully mastered, the company broadened its understanding of its business role by daring to try to see its future beyond the traditional viewpoint, approaching forecasting in a way that it had never done before.

In the six parts of this book, I have adopted a structure that offers readers many ways to address the content at their own pace. The story of Brussels Airlines provides the backbone for a management development proposal. It is systematically synthesized in a 'bear in mind' section in each chapter, with the addition of some exercises to allow readers to assess their own situation with respect to the theme. But because I am also an academic, my goal is always to address a deeper and broader perspective by confronting the experience and not limiting the discussion to one sector, airlines. In order to broaden the perspective of what I am developing, I invited seven CEOs and chairmen to enter into a dialogue with the reader, commenting in the light of their own experience on this management proposal for turnaround situations and corporate transformations.

These CEOs and chairmen are:

- Richard Lapthorne, Chairman Cable & Wireless plc (www. cw.com), insists in the foreword that balance sheet management remains a foundation of healthy management behavious that cannot be put on hold for any reason.
- Sébastien Bazin (Part I), CEO Colony Capital Europe (www. colonyinc.com/europe), gives his views on confidence building with his team of CEOs involved in the leveraged buyout

operations currently conducted by this private equity fund, which are disruptive and effective in terms of results. In summary: Avoid casting mistakes!

- Jose Zurstrassen (Part II), CEO, Keytrade Bank (www.keytradebank.com), a young Belgian entrepreneur who discusses the common style and sleight of hand between a turnaround manager and an entrepreneur, and how both create the correct context to give people the opportunity to work to the best of their abilities. In summary: Through closeness to committed people in the middle of the arena, CEOs anticipate the next evolution of their organization to perpetuate the momentum for success.

- Pierre Henry (Part III), CEO Sodexho Pass International (www.sodexho.com), world joint leader with Accor of the meal and service voucher industry, looks at how in a portfolio of subsidiaries some are always underperforming and how the tools discussed in *Out of the Ashes* prevent the development of more structural concerns. In summary: Paying lip service to your business model is an unambiguous symptom that complacency and arrogance are developing in your organization.

- Arthur Lok Jack (Parts IV and V), Chairman of Neal & Massy Holdings (www.neal-and-massy.com) and Guardian Holdings, a genuinely successful entrepreneur from an emerging country, shares his own experience of managing performance drivers to focus the talent of his companies on one objective: affordability for the customer. In summary: Measure what makes sense in terms of management reaction; being technically correct is not a guarantee of superior performance.

- Magnus Welander (Part VI), CEO of Envirotainer (www.envirotainer.com) and CEO and President of Thulé Automotive, explains how forecasting 30 months ahead is becoming a mandatory skill in business. In summary: Forecasting is an advanced symptom of corporate health recovery, so don't fool yourself trying to play with it.

- Bernard Charlès, CEO Dassault Systèmes (www.3ds.com), in the afterword considers turnarounds from the perspective of the best in class who have never encountered one but know what it is, and pays surcere respect to those who have been able to recover their health. That is why, based on his experience and own beliefs, he shares three pieces of advice: don't let your structure spoil your turnaround tempo; keep intact your entrepreneurial spirit; and keep arming at dominating your business sector so that the lessons & the past remain a high-return investment.

Finally, in the short epilogue to this story, I wanted to continue the journey into the day-to-day detail in order to raise a final question: When an organization has been able to develop a new corporate pulse and become skilled at delivering insightful forecasts, can we conclude that the organization is back on the track to healthy growth again, whatever the fierceness of the competitive environment? Certainly! But, borrowing the words of Andy Grove that 'only the paranoid survive', we must also contemplate the true nature of what has been implemented – is it a systematic process for returning to healthy growth or just a remission?

SN Flight 2113 is now taxiing, ready for take-off – so let's go!

THE FLIGHT OF THE PHOENIX

Talk to Sabena veterans today, and many recall dreading bank-ruptcy, way back in 1983. It wasn't till 2001 that financial ruin finally overcame the Belgian national carrier. On its ashes was born SN Brussels Airways, which flapped its wings, but right off looked like a Sabena replay.

Peter Davies joined the following year as CEO. Davies, a Welsh-man, rugby back row and ex-COO at DHL International, Ameri-cas, inherited a staff of 2000 (down from Sabena's 12 000), and a fleet halved to 38 airplanes (three Airbus 319s, three Airbus 330s and 32 Avro 100-seater regional jets). Not exactly a smooth sail. SN's unique situation – part startup, part turnaround, part organic growth (maybe 'start-through' would fit) – called for tactics and reflexes for all four scenarios. It required considerable people-man-agement skills. It demanded a vision of a changing cutthroat sector whose playing field was littered with dead fleets – Swissair United, U.S. Airways and France's Air Lib among them.

In 2002, with competition fierce and cash draining, many laid bets that the end of the runway was near. 'For the first three months at least, everyone probably thought, "What's he doing?"' Davies recalls today.

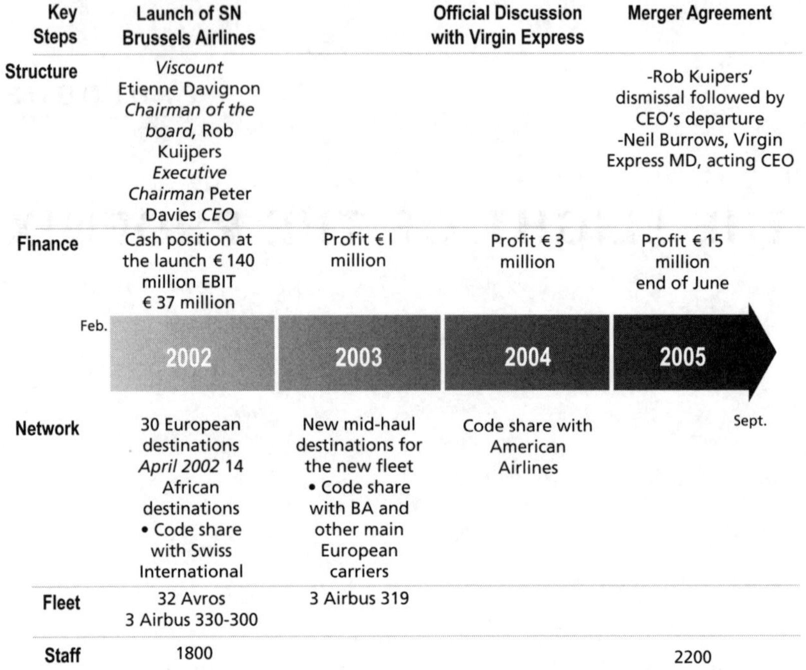

Key Steps	Launch of SN Brussels Airlines		Official Discussion with Virgin Express	Merger Agreement
Structure	*Viscount* Etienne Davignon *Chairman of the board,* Rob Kuijpers *Executive Chairman* Peter Davies *CEO*			-Rob Kuipers' dismissal followed by CEO's departure -Neil Burrows, Virgin Express MD, acting CEO
Finance	Cash position at the launch € 140 million EBIT € 37 million	Profit € I million	Profit € 3 million	Profit € 15 million end of June
	Feb.			
	2002	**2003**	**2004**	**2005** Sept.
Network	30 European destinations *April 2002* 14 African destinations • Code share with Swiss International	New mid-haul destinations for the new fleet • Code share with BA and other main European carriers	Code share with American Airlines	
Fleet	32 Avros 3 Airbus 330-300	3 Airbus 319		
Staff	1800			2200

Figure 1   Overview of the first 43 months of SN Brussels Airlines.

These are the words with which journalist Joshua Jampol introduced the story of SN Brussels Airlines in 'Flight of the phoenix', an article published in June 2005 in *European Business*. SN does not sound like a company that was bound for success, but the key question along this management journey will be continually to assess whether the conditions for healthy growth had in fact been achieved three years down the road in mid-September 2005, when both the executive Chairman and the CEO left the organization.

PART I

THE CEO, THE HINGE OF THE CORPORATE TRUST- AND CONFIDENCE-BUILDING MECHANISM

Without trust, even insightful vision and superb execution are not enough. Trust is a prerequisite in a turnaround or unusual management situation like the relaunch of a company. SN Brussels Airline was one of those cases.

Sébastien Bazin, CEO of Colony Capital Europe, was even more specific in this respect when he explained to me during an interview in early 2006:

> The confidence scheme is taken for granted with the board and results from our collective decision to appoint somebody for this mission. Our responsibility is first to avoid any casting mistake and secondly not to fool ourselves with the nature of the requested role for the mission. For instance, in the Accor case, which made some

noise in fall 2005,[2] the question was to find a person with the right profile to rejuvenate the corporate direction and not keep restructuring. The CEO who was stepping down had done the job up to 80–85 per cent, but repeating the same pattern in this direction would have been a case of suboptimization. So, Accor needed a direction provider and implementer. Gilles Pélisson had this profile and to a certain extent the associated legitimacy, thanks to his experience at achieving this before. But then comes the next crucial question: how this trust contract with the board is shared with the troops. The next day after his appointment, even though I don't like the medium too much because of its sterile nature, he organized a two-hour videoconference presentation with 6000 people all over the world to explain what he was there for and how he planned to achieve it with them. I heard something like 100 comments saying 'He cares about us!' The battle is not won, but for a stakeholder like Colony we can observe that it started on the correct foot and clearly we cannot expect more at this early stage.

In trust and confidence building, the reflex is mainly to focus downwards in the organization by asking the question: Which are the efficient trust-building mechanisms developed by this new CEO? How do the troops react? This is only part of the question, however. In a turnaround context as in prosperity, it is key to observe that trust and confidence are appropriately enforced, in a sustainable way in both directions, upwards to the board and downwards to the staff. This was one of the salient characteristics of SN's initial turnaround step.

In this first part, I will follow the trust-building steps in SN's case and observe their management value in the light of the company's specific context. Secondly, I propose some caveats that

[2] The appointment of a new CEO of Accor Hotels and Resorts turned into a confrontation between board members, with even a disclosure in the French press of who had been shortlisted, leading one of the candidates to be thanked by his current employer for 'unethical' behaviour.

have to be recognized and avoided to secure the desired organizational result. But trust is not permanent. Its foundation can be rocked by corporate events. It is interesting to assess what of the initial trust 'contract' remains workable and what was done effectively in this instance to perpetuate its impact.

CHAPTER 1

DRIVEN BY VALUE

TRUST AND CONFIDENCE BUILDING: A VALUE-DRIVEN MECHANISM

A turnaround must operate as a disruptive process from the company's previous history, otherwise it merely painfully prolongs the past. A new CEO consciously embodies this challenge and this is reflected in the system of values that he or she brings to the new mission. The question is not how these values are formulated, but how the CEO intends to use them. This is the kick-off of the confidence- and trust-building process. The style of the new CEO determines how enthusiasm and passion for the recovery process are ignited. Salespeople say that you only have one opportunity to make a good first impression; the CEO's challenge is the same.

In SN Brussels Airlines' case there were three aspects to enforcing the new set of values: how to be transparent with these values in opening up part of one's private thoughts; how some disruptive behaviours prolong the initial momentum; and finally, how the impact of the values can be perpetuated.

TRUST COMES WITH TRUTH

Back in summer 2002, Peter Davies faced a very straightforward issue: how to bring together all the inherited talents of SN to achieve the initial simple goal of *survival*. But you can't just *decide* to foster energy or talent, this is a consequence of a trust-building mechanism. The birth of the company mobilized some substantial resources among the Belgian financial community. In simple terms, SN's cash in February 2002 amounted to €140 million. The rate at which the company was burning cash didn't offer it a very long lifespan without encountering some crucial difficulties, unless it was able to regain a self-supporting level of operations quickly.

Incidentally, the Swiss International Airlines story illustrates precisely the realization of this risk. Moritz Suter, the founder of Crossair (see www.wikipedia.org/wiki/Crossair), very clearly explained the nature of this narrow path for survival when he said:

> They (SN) began modestly based on reality and true market needs. While they built up operations and made a profit, Swiss annihilated €2,5 billion and was bought by Lufthansa. I am pleased that decisions in Belgium are taken more intelligently.

Air France's Director General, Patrick Alexandre, echoed virtually the same opinion:

> SN Brussels did something radically different from SABENA in terms of ambitions. In fact they managed their risk appropriately, while Swiss replicated Swissair's ambitions and its failing model, with just a difference in scale.

Peter Davies, whom I had known since his DHL days, invited me in July 2002 to help him design the first management conference that he was planning for early September, and to contribute as outside guest speaker and moderator. One of the goals for this first large-scale management development event was to instil a strong sense of urgency in the team, whatever encouraging signals had been received so far – for instance code share agreements, or the withdrawal of some opportunistic newcomers on routes historically operated by SN such as Newcastle or Bologna.

This does not contradict Sébastien Bazin's observation on page 2 about how fast you need to share the tempo of the mission; Peter's case was slightly different. He had to share the shoes of Rob Kurjpers, the Executive Chairman who had launched the process a couple of months earlier. He therefore inherited a context whose tempo was already partly set. In this situation, his sleight of hand paradoxically consisted in the smoothness with which he took the baton, so that the company only perceived the boost in the turnaround programme due to his arrival.

> **Being new is an advantage: it means you enjoy the benefit of the doubt, which can be significant.**

Secondly, Peter had always expressed sincere compassion over what his new team had experienced through the bankruptcy. So he did not hide the fact that the relaunch deserved some genuine respect. He said that one of his missions, as newly appointed CEO, was 'to tick the box of how to transform this bankruptcy trauma into a winning collective dynamic'. In attempting this goal, he had the advantage of being a company outsider, someone who was neither from Belgium nor from SABENA. He was not linked to the past nor to the decisions that had led to the company's fatal decline. But simultaneously, given that Peter had not hired the current staff, there was no natural reason for them to trust him on the journey back to prosperity.

In this context, it is wise to bear in mind that realistically one must count on the benefit of the doubt and no more. Nationality is not a determining factor in success or failure. The key is results. In Belgium, like in any other country, business results – progress – can obviously be quantified. Numbers are universal and have the same value. Changing a company happens from the inside if the CEO can clearly explain what is needed and the desired outcome.

For the first management conference the question was which theme to start with. Peter and I strongly believed that although SN Brussels Airlines was in the airline sector, it was just a business like any other. If we had been in another sector in the same context, we would have tried to start exactly in the same way; that is, on the correct foot. Good practice is correlated to the level of margin in one's own sector. If we were in the automotive industry, higher margins would correspond to the lowest level of defects per vehicle; Toyota is a good illustration of this principle. In an airline, higher margins have always correlated with punctuality.

So there we had a theme for the management conference: punctuality. By chance this was neither fiction nor a gimmick: since its relaunch SN Brussels Airlines had ranked in the top two in Europe in terms of punctuality. So the theme of the conference became: 'Invent a corporate tempo to be the most punctual and the most profitable'. The rationale was, in order for it to be observable by the passengers:

- First, punctuality must be verified by a tangible and unquestionable fact, such as a ranking achieved by an independent body.
- Secondly, the passenger's perception is also influenced by the attitude to punctuality and the practices followed throughout the organization, down to the smallest detail of operations.

Summer 2002's results were very encouraging in terms of ranking, but this did not mean that they reflected a superior level of organizational effectiveness, which would have been quite unrealistic less than one year after bankruptcy. Realistically, the low traffic in Brussels airport after SABENA's crash substantially explained the results. But the results were still good news. Being able to harness the energy created by good news is a characteristic of a successful turnaround manager. Therefore Peter's challenge was very basic: since SN had achieved this level of performance, it could not be wasted and collectively this coveted advantage must be capitalized on.

A good theme was part of the challenge of bringing everyone together, but it was not enough. To counter people's natural scepticism, Peter thought that it was important to explain what the common challenge was with something quite personal, such as an icebreaker. His idea was to share with his management team of 40 senior managers some lessons from his great uncle, who, he said, influenced the way he manages his life. (See Business story 1.1.)

Business story 1.1: A mirror for being at ease with oneself

My great uncle was in the SAS in France and Belgium during World War II. Very early on, he influenced the way I should design my life, my career. One of his most salient lessons regarded these who must live with compromises and these who have chosen not to. Many times in his farm in South Devon he hammered into me, if you want to make your living in politics, you need to be ready to be commanded by hypocrisy with compromises . . . I think that he had some unpleasant experiences in this respect. But due to his senior officer reserve, I could never dig further.

Growing up, I understood that the priority was not the superior nature of the drivers you have chosen for your life, but your consciousness of them, how you live with them. So, I came very early on to the conclusion that the less conscious you are of them, the more energy is wasted in developing justifications. The reason is simple. You are lying to yourself! The paroxysm of waste or inefficiency is reached when you know that these justifications are wrong. This raises a crucial question: are you conscious of what you are doing? This also means, aren't you lying to yourself? This inevitably leads to lying to your collaborators, your customers, your shareholders. It is what I want to tell people first: a turnaround starts by not lying to oneself. We are all in the same boat, relying on each other's honesty, the commitment of the person next to us.

But I also have the antidote. The best repellent I have discovered to protect truth is a simple tool: a mirror. Not Snow White's sophisticated one, there is no question and answer, but a very straightforward form of introspection. Without any form of sexual perversion, can you look at yourself in the mirror, in the eyes, naked without any mask, so without any way to avoid answering the question: 'Am I telling the truth or not?'

After this decision to use a very personal story to support the basic value of truth in building trust, Peter carried on commenting on the rationale for the approach. In his opinion, the answer you give in front of the mirror is crucial. 'Truth as a central personal value conveys one's own long-term consistency,' he commented. 'Over time, it becomes one's own most valuable ally.' We can easily follow Peter's principle and agree that it represents not only a very reliable asset, but also a prerequisite in the demanding challenge of organizational turnaround. The turnaround environment is a relentless one where there are no excuses. The principle 'don't make the same mistake twice', which is heard too often as a way to describe a learning organization, is not the right behaviour in this context. From the beginning, 'do it right first time' is a mantra that any turnaround CEO must impose on themselves, because time is the scarcest resource in this type of challenge. Consequently, starting with the right individual attitude is a must.

The anticipated output of the Limelette[3] seminar was clear: develop truth-based trust as the common platform on which collective efforts can find their meaning to give a future to this company. While SABENA's bankruptcy was a dramatic event, the relaunch out of its ashes was a team achievement (involving first the company's designer Etienne Davignon, who quickly set up the new team at SN Brussels Airlines). This kind of context is a superb opportunity for a CEO to express the concept that for the team's talent to be turned into business success, everyone must be at ease with themselves in order to be at ease with others. This observation echoed Jack Welch's opinion, quoted by Geoffrey Colvin, *Fortune* senior editor,[4] 'You have to be comfortable with yourself to be a good boss.' This is true for any manager and in any situation – for being a good father or mother, a good husband or wife, the rule is the same. In summary, this aptitude for being at ease with oneself is particularly critical in a turnaround, because it helps create a silver lining to boost recovery. And it spreads like wildfire!

DISRUPTIVE BEHAVIOUR PROLONGS THE MOMENTUM

Stories like that of the mirror, and the values associated with them, operate like a milestone in a CEO's relationship with their senior management team. The question is how the impact of reaching

[3] Limelette was the name used in the company to remind people what occurred during this first management convention, which was held in a castle transformed into a convention resort to the south of Brussels.
[4] Geoffrey Colvin (2006) 'Catch a rising star', *Fortune*, 30 January.

this milestone can be prolonged. An effective CEO relies on a library of milestones, which represent the fuel for building closeness with the senior management team. But the art is also to find a means to prevent key values becoming commodities. Disruptive behaviour can be a way of refocusing attention on some key values (see Business story 1.2).

Business story 1.2: A chair and two flipcharts

At the end of summer 2004 the trend of improved operating performance was still delivering encouraging results at SN. But a major event occurred during the summer: the fuel price rocketed. August is also SN's lowest month of activity. One of the challenges consisted in achieving an August where the sparse traffic was well controlled by strictly assessing operational needs. It was better to cancel flights than to fly with poor forecasted traffic.

One exercise, carried out months in advance, was a simple but very healthy break-even analysis. One of the key components of the variable part of the direct operating cost (DoC) of a flight was fuel, which represented around 17 per cent of the cost of SN's European operations. In August 2004 the fuel price was under strong pressure. Moreover, the possibility of imposing a fuel surcharge was still very limited. Finally, even though the planning was correct, the fine-tuning of the final step of execution, which transforms a mediocre performance into a superior one, was missed. Therefore, the loss in August 2004, which was supposed to be half the previous year's, exceeded it by 30 per cent. This represented a major concern, because after eight months of effort, this one bad month erased all the contributions that had been accumulated since the beginning of the year.

Senior management needed to get together to explore the nature of the new challenge. Again, what should be the theme of the meeting? This question was at the heart of the team's commitment challenge. The cause of the low performance was clear, but we could have run the risk of hearing 'It is those arrogant guys in planning who have not done their job' with the easy reply 'Sales did not make enough effort to boost the number of passengers on the planes'. But all this was nonsense, because knowledge of SN's traffic patterns told us that these 'what-ifs' would not have had an impact. Moreover, we could not organize a management conference on slashing costs because that had been the theme of the previous May's meeting. Cost control was the theme, but Peter commented, 'Now we are beyond projects, we end up in the eye of the cyclone if all the senior managers do not understand that this challenge is not limited to a fuel-hedging one.' The airlines have the sad privilege of not making money or very little, while the rest of their sector is coining it in. Many of their costs are imposed and consequently not negotiable, so the answer to success lies in quality of management and responsiveness.

Peter therefore decided to run a one-day management conference with 60 of his senior executives next to headquarters. The senior managers were quite annoyed because they did not receive an agenda. Nadine, the CEO's personal assistant, was asked 20 times on the day before the meeting, 'This is weird, what does it mean?' The team was merely expected for breakfast, 30 minutes before the beginning of the conference. Moreover, when they arrived in the conference room the setting was not the traditional one, just one chair and two flipcharts in front of the audience. The conference turned into a one-day transparent dialogue on how the future was influenced by superior operating results achieved in one's own department. Peter was alone on a foot-high platform so that he could see all the audience even when he was sitting down, and that was how he ran this unusual management conference aiming at reinforcing peer accountability.

The comments I heard indicated that this disruptive approach not only gained the senior management's attention, but also gave a new tempo to rejuvenate the momentum of the turnaround. The meeting was followed by a dinner. The head of ground operations, also called Peter, was sitting next to me and outlined a discussion he was then having with the government regarding the pilots' pension scheme, which cost €4 to 5 million a year. As a result of this session, he began to question his own impact on the company's overall cost management performance, and concluded, 'If we don't obtain renewal of the "Arrêté Royal" [a legal ruling regarding the pension scheme] SN loses 1.5 points of RoDoC[5] when one estimates the current RoDoC performance on the European network at around 105. This is nearly one third of the fleet contribution . . . this session was a powerful reminder to put my efforts into perspective.'

This session worked well, but a chair and two flipcharts formed the visible part of a more complex iceberg, which needs to be briefly described to provide understanding of the whole dynamic. Returning to key values demands that the CEO retains a fair sense of balance and undertakes strict preparation within a very short timeframe. In SN's context, the fair balance concerned the management of urgency and hope. Peter enjoyed a great opportunity to boost hope and company confidence by unveiling an advertisement that had just been created. This advertisement was shot with 200 collaborators who volunteered to spend a full day on the tarmac of Brussels airport, generously getting sunstroke and aching legs running after an Airbus 319. In terms of team building, hope, never-been-done-before style and also cost management, this ad was a winner. It even got an award and, before being broadcast on Belgian television, was seen on the internet by 250 000 people.

---

[5] RoDoC stands for return on direct operating costs. This performance measurement concept is extensively developed in Parts IV and V.

Once the hope had become tangible, the profit-and-loss side became a mere question of discipline in preparation, because such an exercise cannot support any form of improvisation. A CEO works without a safety net and senior managers can very easily spot any approximation. Publicly questioning a department's results and linking them with operating performance at the elementary level of strategic execution[6] is ambitious, but the return is worth the challenge. Do we keep telling ourselves the truth about the performance of our own department or are we unwittingly letting complacency develop? A strong value-driven message and a disruptive communication style build confidence and closeness with one's own management, but this mechanism demands even more effort to perpetuate its momentum.

PERPETUATE VALUES THROUGH SYMBOLS

I warmly encouraged Peter's choice at Limelette to strike his senior executives' minds with the mirror story. But my real concern regarding the effectiveness of the message in such a context was how many days after the management conference Peter's principle of trust through truth would start to fade. Under the pressure of events or the daily routine, how quickly would the team start making the wrong trade-offs between what they had heard and how they were prepared to behave? Instilling a new set of shared values is a mandatory step in the context of a

[6] The concept of elementary level of execution pushed one step further the granularity of the concept of a business unit attached to portfolio management or strategic planning, to be sure that one monitors performance at the correct place so that the management reaction can occur quickly and be appropriate. In the case of an airline, the route seems the appropriate place; in a distribution chain it would be the outlet. The goal is to deal with what is managerially relevant; what is technically correct is insufficient.

start-through process, but the values to be shared with a team of senior managers and consequently with the whole company are not sustainable by themselves. They need to be nurtured for them to last.

I was not questioning the relevance of the mirror message, but reacting like any academic in front of such a brilliant idea. I was thinking in terms of assimilation, the ability to apply the message appropriately in one's own environment. Would this mirror story, selected to build the foundation of a new management attitude and style, be enough to secure the intended transformation in behaviour? We have all experienced this situation: leaving a session enthusiastic, as lecturer or participant, but eventually encountering disastrous results at the next test on the topic. The reason is simple: enthusiasm is the icing on the cake of the assimilation process, not the cake stand. So the question was, how could we prolong the impact of Peter's basic principle in the turnaround context?

The best ally for prolonging an impact is to find a symbol of disruptive behaviour. The mind forgets more easily what the eyes can see permanently. Quite often turnaround champions who are achieving their mission have a set of skills where *style* and *shared values*[7] are as important as cost-cutting recipes. One needs the

> **The mind forgets more easily what the eyes can see permanently.**

help of something stronger than words. Symbols are powerful because of their permanent, active presence: they keep reminding us of the key lesson about the style, the attitude with which a company should collectively endeavour to find the path to return to prosperity or to stay prosperous (see Business story 1.3).

---

[7] Style and shared values refer to two of the McKinsey 7S concept, addressed in Thomas J. Peters and Robert H. Waterman (1982) *In Search of Excellence*, New York: Harper & Row.

Business story 1.3: A library of symbols

In 1983, at what became Carnaud Metal Box (CMB) in the late 1980s, I had the opportunity to collaborate on some consulting assignments with the CEO, Jean-Marie Descarpentries,[8] who is an expert in the management of symbols. A decade later, he accepted the challenge to turn around Bull, which had accumulated €3 billion of losses over the four years before he took charge.

The French-based IT company experienced positive financial results under his management. At Bull, as at CMB, he used his sense of symbol management as a powerful tool to first ignite and then accelerate change. Among his early decisions, at least two reinforced the management of values illustrated earlier.

A few days after his arrival, he left his presidential office on the top floor of one of the towers of La Défense in Paris and installed his office on the salesforce's open-plan floor, right in the middle of the stream of information regarding the company's survival. 'I needed to plug myself into the business reality to save time. Reports are fine but I need tangible facts,' he commented to his troops and added, 'I am just starting . . .'

A few weeks later, considering that this address in a steel, glass and marble tower and its associated costs were not consistent with the company's current financial situation, he decided to relocate the headquarters. 'There are always idle assets that can be perfectly useful if you break some mental representations of what is and what is not appropriate,' he said. Some

---

[8] Jean-Marie Descarpentries is the former CEO of Carnaud Metal Box, a group that he contributed to developing and transforming in the mid-1980s. He became the CEO of Bull in the mid-1990s, a period during which this company nearly knew its only period of profits.

ex-NATO premises in Louveciennes near Versailles were a good symbol to remind people that the company couldn't burn cash it hadn't got. In order to save money both by reducing the rent and downsizing the functional staff for obvious physical reasons, there was limited space at the new premises. The message in terms of rationalizing costs was unambiguous. But the move was also full of surprises, one of which was very relevant to SN. When Jean-Marie arrived in what was initially designed as a barracks, he asked that the door of his office be removed and left in the corridor next to his executive PA's office. 'My open-door policy is not an accident. The signal that it is a deliberate management value must be clear and permanently understood as such over time,' he commented.

After Limelette, a mirror was planned to be installed on the right-hand side of the door of the CEO's office when SN moved to its new facilities, gathering together all the departments scattered around Brussels Airport.

CEOs who are leading a turnaround must do it with a strong management philosophy, principles and values, supported by symbols to perpetuate their impact. But it is necessary to add a nuance, which I heard in a discussion with Xavier Huillard, CEO of Vinci the world's leading French-based civil engineering and concession group (www.vinci.com). Xavier Huillard's comments on the management of corporate values help remind us of the constant challenge of optimizing the impact of those values. His opinion is:

> Values are crucial, but as soon as you write them down you run the risk of falling into the trap of interpretations and comments. So if I can avoid it, I don't carve values in stone, reserving the opportunity to fine-tune them continually as the context evolves. But not writing them down paradoxically imposes a more demanding discipline. One must hammer in the value message again and again at any opportunity for contact with the collaborators, the customers,

the board members, to scrupulously manage the consistency of its evolution and keep all the stakeholders aligned on the company's ultimate goal . . . its survival and its independence.

Turnaround is about turning the page from the past. The management style of the new CEO must reflect this willingness. Finding the correct tempo for the new era is one thing, but it must also be supported by active reminders, which is the role of symbols. Moreover, the search for the correct symbol is not a management trick but a mandatory step in securing a transformation process. Finally, transformation is about continuous adjustment. It is a mistake, or more precisely a form of complacency, to formalize the transformation into a glossy paper document, which by its intrinsic nature limits the opportunity for micro adjustments to fine-tune the effectiveness and the relevance of the transformation process. So the CEO must welcome any contact with all the stakeholders – staff, board members, customers – as an opportunity to repeat, repeat and repeat again the fundamentals of this transformation. In the first steps of the turnaround this is a job for the CEO, but they must be quickly joined in this exercise by their closest collaborators, to create a critical mass of transformation in the whole organization as quickly as possible.

CHAPTER 2

IT WORKS BOTH WAYS

BUILDING TRUST AND CONFIDENCE OPERATES IN BOTH DIRECTIONS, UPWARDS AND DOWNWARDS

Mechanisms for building trust upwards and downwards are of equal importance. In the previous chapter, I discussed the keys used by the new CEO of SN Brussels Airlines to spark the downward trust- and confidence-building challenge effectively and head towards some tangible results. But this does not mean that the upward side is simultaneously guaranteed. The results cannot be taken for granted. That is why in this upward trust-building mechanism one needs to rely on genuine evidence, which takes us beyond the enthusiasm of any initial story.

A BOARD AND A CEO ON THE SAME WAVELENGTH

Being so much absorbed in engineering the turnaround at SN, I nearly focused all my attention on how the CEO was consolidating trust downwards. So I took for granted the trust between Peter Davies and the Executive Chairman (Rob Kuijpers) on one side and the President of the board on the other side.

Peter did not get the job as CEO by accident, as was suggested by some comments in the Belgian press[9] at the end of team Kuijpers–Davies in late September 2005. The reality was completely different. This well-documented article just drew a few pictures, while a movie would have been a more appropriate way to get a sense of the turnaround.

The logic behind Peter's recruitment was simple. When Rob Kuijpers[10] left DHL, a multilingual high-calibre top manager with an impressive track record of senior leadership positions became available in Belgium. SABENA's bankruptcy occurred almost simultaneously. The 'father' of the relaunch of the Belgian national carrier, Etienne Davignon, quickly saw in Rob the required talent and partner for such a mission. But in 2001, Davignon and Kuijpers were respectively over 70 and 65 years old and the relaunch – borrowing the words of Joshua Jampol,[11] the *start-through* – could not be executed by either of these two prominent leaders. They could influence, give direction, moderate and encourage, but they could not act as first in line. So a star had to be found for the execution.

Secondly, Rob did not find the necessary turnaround talent among his two or three closest senior executives, available out of the ashes of SABENA. Quite naturally, he looked in the direction of the men with whom he had built his current reputation. At this stage and in this context, whoever Rob had found to run the

---

[9] 'How Davignon ousted SN Airlines Executive Chairman', *Trends*, 19 September 2006.

[10] The success of Rob Kuijpers at DHL must not lead us to forget that before his success period, he was involved in a couple of critical turnaround examples with both DHL and Heinz.

[11] 'Flight of the phoenix', *European Business*, June 2005.

execution side of the turnaround would have been seen as the right person because he or she was the white knight who had the courage to take up this challenge. Moreover, this was not a simple recruitment exercise. SN, by its nature in 2002, could have imploded within six months. In this sort of case, only personal networks are effective to involve someone quickly who can jump almost immediately into the heart of the arena. Consequently, any time-saving solution was welcome and appropriate. So it is fair to conclude that Peter was Rob's best possible asset at that time.

Business story 2.1: A necessary evil or the right person?

At the end of DHL Americas country managers' convention in Miami in mid-February 2002 – where I ran a very challenging workshop called 'Killing your Ambassadors', based on real experiences of service delivery and recovery failure – Peter and I were waiting in the lobby of the Biltmore in Coral Gabble for an airport transfer. We were talking about what each of us was doing the following weekend. Peter said casually, 'I've been asked to consider a CEO position.' We kept talking, but this was not breaking news for me, since I knew Peter's market visibility. So I did not question him much about this company, which sounded completely unknown. For me, it was just another potential challenge for Peter, in line with the internal one he had already taken up in DHL. Moreover, the airline sector is a business like any other, particularly in turnaround conditions. In this context, the core skills are managerial and sector experience is not so crucial.

Arnaud Fayet, a former executive committee member of Wendel Investissemenents (www.wendel-investissement.com) who has personally experienced this kind of challenge and who also worked closely with Jean-Marie Descarpentries, reinforced the above, saying, 'In the Bull case, Descarpentries had applied the lessons from his CMB experience. After three years, he stepped down because his period of highest value delivery was over and, among many other reasons, he was wise enough to recognize that the company was entering again into a phase where technology choices were crucial. In this context, the generalist must pass the baton to the sector specialist.' Alan Lafley said exactly the same, commenting on Wall Street's reaction after his appointment as CEO and Chairman of Procter & Gamble, 'P&G lost some points. The appointment of a sector specialist does not usually mean spectacular change, the first flavour is more about organic growth and hard work. So no spectacular announcement.'

Peter and I kept talking until we reached the gate for the Paris flight, where I left and, just in case, suggested that Peter should think about the answer to this question: 'For your CEO position, is your future employer considering you as a resource or as the right man? If you are lucky and it is the latter, please check which characteristics of your skills and style they are assessing as differentiating and critical for the mission. Be sure that there is a shared understanding of your talents!'

Three months later, in Zaventem at a typical Belgian auberge near the airport, I had lunch with Peter, who had just joined SN Brussels Airlines as the new CEO. Retrospectively, I remembered not having received his answer to the question I'd posed him at our previous meeting (see Business story 2.1). Subconsciously, I thought that the trust-building issue was just a downward one, because Peter's new boss was his former one and the genuine reasons for his recruitment were obvious because of their common track record.

> Why did the board offer you a CEO position? Skip the question and run the risk of a casting mistake.

Know the real reasons for recruiting someone as CEO. This deserves some extra investigation in order to be sure that both the CEO and the board start on the same footing. There is no shame in investigating the rationale for one's own selection. I observed a similar situation very recently and the results are worth the effort.

CMD, one of the companies used to support the concepts and principles of my previous book, *Organic Growth,* was managed by a brilliant, insightful, hands-on, young CEO. First he turned around the company in the mid-1990s, then he consolidated it and led the effort to find an exit solution when CMD's owner needed some cash. But the sale of the company to Textron in 1999 was aborted for some completely ill-founded reasons. So the CEO kept improving CMD's cash position and profitability, until the owner asked him to extinguish a serious fire elsewhere in his holding company. He turned around this basic chemical operation over 18 months, but meanwhile the mechanical company was back under the spotlight. One of its competitors, specializing like CMD in customized heavy mechanical products, was very eager to buy CMD in order to consolidate the world leadership of both players in some of their expertise areas. But the deal was subject to a break-out clause, that the former CEO returned.

I had lunch in December 2005 with this CEO to discuss some aspects of this book just after the acquisition deal was closed and I asked the same question as I had posed to Peter: 'What does the future owner – a woman, the daughter of the founder of the company – want from you as CEO? Which skills and style is she basing her decision on?'

Here again, this was an issue of map reading, which lies at the root of whether trust can be expected to be built upwards or not. In this respect, confusion, half-truths and interpretations are traps. Only one rule can avoid costly frustrations: call a spade a spade. This represents a good test to assess whether a company, and especially its board, is both capitalizing on its current strengths and not remaining exposed to its current weaknesses. One form of this threat is to prolong the past, as Sébastien Bazin commented in the Accor case (see page 2). The question is not the skills of the potential CEO, but their appropriateness for the current challenge as defined and agreed by the members of the board. Unfortunately we think too often by analogy rather than assessing the perfect match in detail and with insight. I suggest that future CEOs should actively commit themselves to this assessment. Casting mistakes are too costly. That is why it is worth scrupulously deploying a superior sense of accuracy in surveying the landscape through a sensible form of map reading.

At our first meeting in Brussels, I did not check with Peter whether or not he had validated the real reasons for his final selection, although I should have insisted. The question of the reason for the choice was also prolonging the relationship between Peter and his former-and-future boss, Rob. Peter achieved outstanding performance at DHL. He turned around the UK operations, pioneered business in the Mediterranean Rim[12] and rejuvenated the company in the Americas. He did that with total empowerment from Rob and simultaneously full accountability from Peter, but at a distance of 6000 km. Now, the game was different, they were only 6 metres apart. Consequently, two instances of trust building with the same characters could have different flavours. The context defines the interaction between people, not only the individuals themselves. There is therefore a high risk in sticking to the prin-

[12] An area including the eastern part of Europe, Russia, Turkey and Israel.

ciple that one does not change a winning team. This does not mean that the team must necessarily be changed, but that one must take sufficient care to consider how the new nature of the game could affect the use of the critical skills that were the foundation of the previous success. Hopefully in this case, the initial winning formula would deliver its foreseen results.

Peter was now in charge as CEO and it was legitimate for him to consider that upward trust was taken for granted due to the context. Consequently, he could dedicate his entire energy with passion and commitment to building downward trust and fulfil the three objectives that he and Rob had promised the board:

1. Give the heart of Europe, Brussels, a 360° carrier.
2. Rebuild the confidence of the staff.
3. Make a profit.

In the previous chapter, we saw how the CEO's value system ignited this trust-building dynamic. Now I propose to dive to a deeper level of detail and describe how this process continually faces the risk of deviation from plan and the level of scrupulous care required to reach the intended results.

BUILDING TRUST AND CONFIDENCE DOWNWARDS, BEYOND ENTHUSIASM AND TALENT, THROUGH DISCIPLINE

The SN case highlights the importance of the CEO's own corporate discipline in managing this process in three directions that reinforce each other. First, any deviants, those who by their attitudes and behaviour might generate doubt or confusion about the objective pursued by the CEO, must be brought onto the correct track without any delay. Secondly, the new CEO inherits a structure that they have not designed and that can contain some traps, therefore it is at least wise to mark them on the pitch. Finally, trust building is a collective corporate game, leveraging any good team news or individual performance to cement the collective dimension of the recovery.

Discipline to constrain any deviants

Downward trust building is a demanding process, the detail of which must be continually controlled. This means strict discipline on the part of the CEO not only to prevent the occurrence of deviant behaviour, but also to correct it immediately it occurs. Some managers are more used to trust building thanks to their leadership or charisma, but this is not a reason for them to be complacent and tolerate any form of suboptimization coming from the attitude developed by a minority, even in some cases a single person.

Building trust downwards is a question of building context. This context cannot run the risk of being spoiled by some deviant attitudes, which can spread like viruses. Beyond the principles, the solution for any manager as well as a CEO is again a map-reading exercise. This consists in plotting the nature of the behaviour of one's closest collaborators to assess which game they are ready to play. Can one count on them or not? The matrix in Figure 2.1, borrowed from *Organic Growth*, offers a quick acid test to prevent a waste of the leaders' energy.

A turnaround challenge is a team game, where cohesion in the group of senior managers is critical. Building confidence is a very fragile process that is sensitive to the smallest details. One cannot run the risk of some behaviours unwittingly being interpreted as a lack of commitment. For instance, during the Limelette seminar, the CEO and one of his executive committee members had not fine-tuned their respective agendas, and this led to unnecessary questions and even gossip by the whole group.

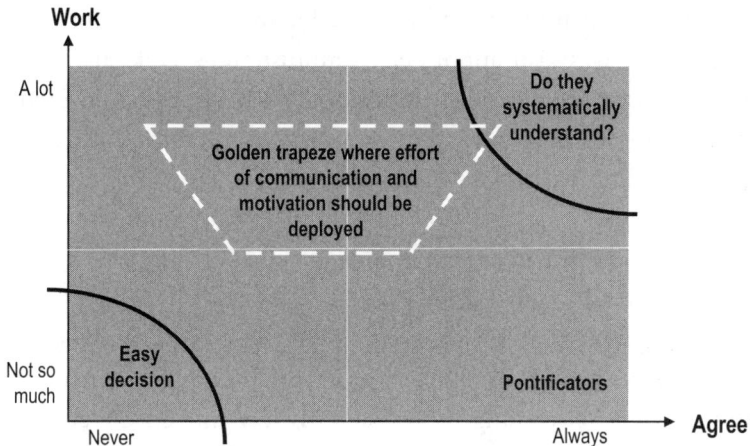

Figure 2.1   Who I am playing with?

The CEO must undertake a careful review of the rules of the game with their senior management team. Just because people are executive committee members does not mean that their team-building strings are already tuned to the right pitch.

Discipline to decode the trap of the inherited structure

The question now is, why do misunderstandings and tensions happen? Is it due to the personality of the people involved, to the structure, to bad luck . . .?

My experience suggests that one needs to pay very careful attention to the structure, especially when one has inherited it. The structure creates the context, which then justifies attitudes and behaviours. In SN's case, the reasons for such frictions developing could be immediately read and anticipated by observing the structure of the organization. For instance, the company's Secretary-General reported to the CEO. But technically, he worked directly with the Executive Chairman and the President of the board. This was a weird situation, which became even more complex in October 2004, when the Secretary-General's function was transformed into that of a Chief Financial Officer and simultaneously he was also appointed Administrateur Délégué[13] of SN Airholding, the legal entity that controls SNBA and Virgin Express. Just by trying to draw a map of these intertwined forms of power, channels of information and relationship systems, a simple question emerges: who is the real boss of whom? Who holds the power? Developing confidence downwards in such a context is always possible, but with what expenditure of energy?

Consequently, at the beginning of the SN story when one tried to draw a map of the upward trust-building mechanism, it was somewhat confusing. So very early on the CEO, conscious of this context, focused all his energy on building trust downwards, in areas where the return on his investment would be optimal for the company. He developed a specific focus in the execution of his mission on protecting his staff from being diverted from the main goal. The CEO's talent also translates into leading people, but leading with subtlety to nurture the employees' efforts. The reason is simple: at some stage one has to stop talking, the facts say whether one is right or not. But the commitment to get the facts right cannot be diverted by polluting factors. So it is the CEO's mission to create this context, which also becomes a real test of their trust- and confidence-building mechanism.

One must stay very vigilant and eradicate anything that can weaken or slow down this process. Deviant attitudes and

---

[13] Administrateur délégué means CEO in Belgian corporate law. So this function means CEO of the holding company that holds the shares of its stakeholders. This has no direct executive impact but a very serious influential one. That did not escaped the magazine *Trend*, which on 22 September wrote an article, 'Wrench in SNBA', in which it said: 'Rob Kuijpers was between the hammer and the block with his management on one side and the holding management personalized by Etienne Davignon. Moreover his Financial Director is also a director of SN Airholding, which creates a complex reporting context. This was enough for the Dutch leader: he moved his office on the spot.'

behaviours constitute one element, but in my opinion the worst threats are always related to structure. They hide like landmines. One can only partly accommodate them in the short term, and that is always shorter than expected. My recommendation is similar to when building confidence upwards. Building trust downwards requires not only mapping the behaviour of different groups of collaborators, especially in the executive committee and among the senior managers, but also achieving a clear reading of the structure to anticipate how it could weaken the CEO's position in case of tension with the board.

Discipline to promote collective performance

Building the confidence of the staff is a never-ending mission. This discipline is a continual requirement and vigilance cannot be taken for granted.

An initial success is not a reason to lessen one's own vigilance, because the foundation of a corporate renewal can also be undermined. The best repellent is the CEO's sharp communication style, making it known that there is only one corporate way to read good news.

The early days of a turnaround are scattered with little aches and pains. But at the operational level, these do not prevent the different managers from restoring confidence day after day in their own performance and consequently in the real potential of their new company. The accumulation of success helps reinforce this opinion. News came at the end of 2003 that brought unexpected encouragement to the whole company. SN Brussels Airlines was ahead of plan, posting its first profit of €1 million after its second year of operations, while simultaneously its former SABENA relative, Swiss International, was reporting disastrous results and still burning cash at a crazy rate. Without making any noise about it, SN was already back into a cash-rebuilding phase. At this stage, the whole company received unquestionable evidence that positive signals were accumulating, and that something unknown in its history was happening. This was based on a clear business reality: the company was gaining passengers on its European network, performance-optimization practices were turning into reflexes, and African operations were fulfilling their anticipated role. The company was making a profit!

This kind of performance is a legitimate reason to thank the whole team for their efforts and prolong the momentum by gathering even more collective energy. This was done through breakfast sessions delivered over three days by the Executive Chairman and the CEO in three languages, sharing and fully discussing where the company was heading with the flight crew and other members of the company. SN was profitable after only two years and this was a hard fact. As I said earlier, SABENA had made a profit only twice in 75 years, once in 1958, the year of the World Exposition in Brussels, and again in 1998, when Swissair played with some accounting principles. So SN Brussels Airlines could be proud to lay some of the ghosts of its history. A page had really been turned and something had to be done to consolidate this feeling. Success is not enough: one must pay attention to any form of disruptive message that can divert the organization from the turnaround's single hymn sheet, as Business story 2.2 illustrates.

Business story 2.2: I manage this company like a hedge fund

'I manage this company like a hedge fund, and one must be clear on the origin of the performance,' SN's Secretary-General used to say. This meant that he saw himself as a substantial contributor to the successful results, which was unquestionable at this stage of the company's history. In 2003 the profits combined a limited dose of operational performance improvement and a lot of exceptional results.

In a turnaround, individual performance supports team performance, which cements the company's transformation. Turnarounds based on individual achievements unfortunately don't last. It is like the two goals scored at the last minute of the Euro 2004 football championships by Zidane against England. The French soccer team was completely lost in terms of direction, but Zidane turned a potential lost game into a last-minute success. This kind of individual achievement is always pleasant to watch on television, but it was not enough to avoid an exit later in the competition. The French team did not work well together; a donkey has never won a horse race. This simple observation is also true in the context of a turnaround. It is a never-ending process of team building, transforming former losers into players.

Individual achievements are very good news, or positive bricks to build one's way out of a crisis. They must not be wasted, they need to be tracked, as Jean-Marie Descarpentries did at Bull with his 'Good News Bulletin'. Good news prolongs the momentum if managed appropriately. But the goal of turnaround engineers and implementers is one step beyond this, as Anne Mulcahy expressed so appropriately (see page XVII).

Putting the spotlight on one's own contribution is not constructive, because it diverts attention from the ultimate goal. Inaccurately assimilating this principle leads one to run the risk of turning what was supposed to be a real turnaround into a mere remission. The CEO and the turnaround team act with an ultimate goal in the back of their mind: to eradicate the cause of the disaster and anticipate elements they don't even know about yet. In his book *The Airline Business*,[14] Professor Doganis quotes the former President of Continental Airlines remarking, 'I have never

---

[14] Rigas Doganis (2005) *The Airline Business*, 2nd edn, London: Routledge.

seen a team which manages a company into crisis to get it back on track.' A turnaround manager cannot afford to be one war too late! Like pilots, turnaround managers must look first in front of them in order to quickly develop some skills to see beyond their field of vision.

Protecting the cohesion of the new company can be illustrated by an iceberg. The positive financial results are the visible part. But in order to achieve the necessary recurrence of positive financial results characterizing a company that is back on a healthy growth track, the turnaround process must be given a management development dimension, which is the portion under the water. This, unfortunately, cannot be taken for granted once the first financial results are obtained.

Business story 2.3: Good news stories are one-offs, bad ones reoccur

In SN's turnaround, the first profit was made from a lot of exceptional results. This was the rule of the game, but it did not make it any less valuable. What was critical was the speed of delivery of these first results. Luckily, the company got better results than forecast thanks to the Secretary-General's talent for gaining public subsidies and having fruitful discussions with SABENA's liquidators. And he deserves real credit for this result.

Nevertheless, a nuance must taken into account in order not to fool oneself. These unquestionable achievements were unfortunately one-off measures. In October 2005, in a discussion with Viscount Davignon, I noted his very relevant formulation: 'Good news consequences are usually one-off, while bad ones are recurrent.' He used this formula to explain that some exceptional results would once again keep supporting the quality of 2005's performance: 'Some million euros resulting from the settlement of an agreement with BAe, the cause of which originated in the SABENA epoch.' But he forgot to mention that the flipside of the good news was already in the system. For instance, the Belgian government, after long discussions with all the members of the airline sector, did not agree to continue with a very favourable regulatory regime regarding pilots' pensions.[15] This decision cost SN some millions of euros, which would erase the impact of the good news in approximately three years.

Therefore, the recurrent impact of bad news becomes a challenge in terms of operational performance improvement. Only confidence-based team cohesion can generate the necessary creativity to constantly boost the performance of the business in order not to lose ground and fall back into Anne Mulcahy's ditch. The focus of the challenge then shifts from a welcome individual performance to a team one, expressed as the skills that need to be rewired to reach sustainable healthy growth. This management development challenge cannot be taken for granted once the first positive financial results are achieved. It is the central theme to make sure that 'one-offs' become recurrent.

[15] See page 14.

CHAPTER 3

TAKE IT STEP BY STEP

MAJOR CORPORATE EVOLUTIONS DETERMINE THE LIFESPAN OF THE TRUST- AND CONFIDENCE-BUILDING PROCESS

Trust- and confidence-building mechanisms require a lot of care and energy to ignite them, but they are simultaneously very sensitive to discord, which can lead them to fade out prematurely. SN illustrates this risk in three intertwined ways. First, the trust-building process is not a 'one size fits all' topic: each specific corporate context requires a tailor-made approach. Secondly, how one approach can be leveraged with the following one is influenced by the role played by the CEO in the design and implementation of the corporate challenge. Finally, divergence between the CEO and the board on strategic evolution negatively affects the downward trust-building mechanism. This leads to a standstill, waiting for a rejuvenation that may not arrive.

ONE SPECIFIC TRUST-BUILDING APPROACH PER SCENARIO

Focusing my efforts on the CEO's downward trust building through implementing a new management toolbox, I nevertheless had the opportunity to observe the sequence of day-to-day events that led to the departure of the Executive Chairman and the CEO. On 29 September 2005, the Belgian magazine *Trends – Tendances'* front-page headline read: 'How Davignon ousted SN Airlines Executive Chairman'. One can also read in this article: 'It is surprising that the team Kuijpers Davies after having been so estimated are now being held up to pubic obloquy . . .' A well-informed journalist explained in detail what had happened since the acquisition of Virgin Express by SN Airholding. But she failed to underline a key point. The managing duo was perfectly in line with the president of the board until the latter, unilaterally, signalled the end of the first half of the game and offered to play a new one with a different theme once Virgin had been acquired.

> **SN's story was in two acts.**

In my opinion there was not one act but two. In terms of trust building, there were also two acts. The first, until the declaration of intent to do something with Virgin Express in March 2004, was prolonged by nearly a year of due diligence and negotiations. The second one began after the closing of the deal (end first quarter 2005), which led to the conclusion mentioned above. My misleading perception of one act was even reinforced by the feeling of inertia characteristic of this acquisition, about which for instance an SN union leader commented, 'The management does not seem to know what will happen.'[16] Initially, I analysed the situation as a continuum, but a discussion with Arnaud Fayet (see page 23) about rewiring a growth context made me realize I was making a mistake. So I

[16] Reported in *La Libre,* 'But what is going on in SN?' 26 May 2005.

began to read the SN case in acts analogous to the Carnaud Metal Box story (see Business story 3.1).

Business story 3.1: Today's success does not guarantee tomorrow's

CMB was a story in three acts. Act I, a turnaround–consolidation phase from 1983 to 1990. Act II, the merger with the British company Metal Box to form CMB (Carnaud Metal Box). Act III, the CEO steps down.

During the first act, Jean-Marie Descarpentries developed a trust-building dynamic in both directions, upwards vis-à-vis the chairman of his main shareholder, CGIP, and downwards with his troops. Descarpentries was hired as the appropriate man for such a situation based on his outstanding track record at glass companies Glaverbel in Belgium and Cristaleria (the Spanish subsidiary of Saint-Gobain). He was a leading representative of the emerging new breed of 1980s turnaround CEOs. In Europe they included Barnevik at Asea and Maucher at Nestlé. During the first act at Carnaud, the trust dynamic was reinforced both ways thanks to the performance improvement.

In Act II, merger and integration, Descarpentries became the CEO of a far larger structure with two main cultures. Paradoxically, this manager who had built his reputation on leading people (including being captain of an airborne troop during the war in Algeria in the early 1960s) experienced his first problems with his own troops. Some questionable decisions, such as a new headquarters in Belgium for symbolic reasons, broke the trust with the French side while simultaneously reinforcing a form of defiance from the British side. This, in addition to some other facts, finally led to doubts being created at the holding level, CGIP, which eventually also broke the trust of the main shareholder. At this stage there was only one possible outcome.

The CMB case raises the question of how the CEO's trust-building mechanism is affected by a major development in the 'contract' with the board due to a radical modification of both the size and the nature of the management challenge. A major acquisition is one of these situations. In the context of a healthy company, this can completely reshuffle the deck. But in SN Brussels Airlines' case, only 30 months after SABENA went bankrupt, the potential upheaval of the acquisition of Virgin Express was even greater. The wounds of the bankruptcy were barely healed when the organization was heading to the hardest management exercise, the integration of two companies.

These two examples underline the complexity of trust building. CMB reminds us to ask whether the new tempo is the right one. SN raises the question of whether one's own talent to adapt to a new tempo and keep playing has not been overestimated.

THE CEO'S ROLE IN MAJOR EVENTS INFLUENCES THE SURVIVAL OF THE INITIAL EFFORTS

Question 1: Who is promoting the next corporate evolution?

Before wondering about the impact on staff morale of announcing an intention to hold talks with the fiercest local 'enemy', the question is what is the role of the CEO in a radical strategic evolution. In Carnaud's case, Descarpentries was the promoter of the merger. I remember that as early as 1984, I had heard some comments from him, who had been appointed CEO of Carnaud two years before, saying that the earlier the better, the company would pass through a major acquisition or merger to consolidate unquestionable European leadership. A very famous aspect of Andy Grove's management style at Intel illustrates the same point.

He asked, 'If we got kicked out and the board brought in a new CEO, what do you think he will do? Well, why shouldn't we walk out of this door, come back and do it ourselves?' In summary, Grove suggests:

- Never delegate the role of devil's advocate about initiatives regarding the strategic future of the company.
- Develop a paranoid style of behaviour for one clear reason: because one must anticipate what has to be done for the good of the company, its staff and its shareholders.

His company was known as a fierce cannibalizer of ideas from other areas. In fact, Grove suggests applying this recipe to one's own management skills to develop the necessary flexibility imposed by the different natures of the management challenges to be faced. That is not to argue that the Kuijpers–Davies team did not develop the appropriate anticipation for the good of the company. If the CEO is at the origin of the strategic shift, considering that it is the best solution for the company's evolution, he or she sells this clear strategic scheme or insight to the board, which may then be confronted by an execution problem due the different nature of the challenges. Then, following Sébastien Bazin's observation in the Accor case (see page 2) regarding the prevention of casting mistakes, the appropriateness of the CEO's profile for the new challenge can be revalidated, and the fundamentals of the initial agreement reconfirmed as still actionable. However, I don't read the Virgin acquisition in the SN case as an Executive Chairman/CEO initiative, which radically modifies the whole dynamic.

Question 2: Is the interpretation of the key corporate challenge sufficiently engaging?

Rob Kuijpers and therefore Peter Davies received from the board a mission in three parts:

1. Connect Brussels.
2. Save 2000 jobs.
3. Make a profit.

Very few people in the early days of the turnaround believed that this challenge could be taken up successfully. 'Kuijpers gives up after, against all the prognosis, he turns SNBA into an unexpected success,' wrote *Trend* in September 2005, perfectly reflecting a common feeling among investors. They had to cope with a form of social responsibility for their own country. Something needed to be done urgently. But there was a happy surprise: the Kuijpers–Davies team generated positive results more quickly than expected. Paradoxically, this early profit defocused the board and its president from the real challenge of the turnaround.

Early in 2004 the goal remained secure, the most sustainable way to cope with the three points of the mission. In my opinion, the President of the board thought that the mission could be considered as over and that it was time to think of something else more ambitious. This was not a radically new idea and it had already been on the agenda before the bankruptcy of SABENA.[17] Nevertheless, the dynamic of success operated as a trust-building booster downwards through the company. The staff were proud to rediscover tangible evidence of their confidence in themselves and in the whole team. The theme of the moment was turnaround through people and the company was really living by its new hymn sheet. Everything was said and done to demonstrate and convince people of the winning difference of SN's dynamic.

Virgin Express was certainly SN's 'enemy' on the Brussels platform because of its very aggressive behaviour on some routes. This was absolutely normal and well accepted by SN, which had this point confirmed in its Executive Customer Survey as early as the end of 2002. But simultaneously, SN could note with satisfac-

---

[17] Guy Vanthemsche (2002) *La Sabena 1923–2001: Des origines au crash*, Brussels: de Boeck, p. 303.

tion that in this context of fierce competition, in the vast majority of the cases it was outperforming Virgin on a like-for-like basis. So Virgin was a competitor, but due to the growing confidence of SN's staff in their own performance, its evolution was virtually disregarded by the large majority. SN's recovering performance and the competitive benchmarking reports were understood by the team as meaning that sooner or later Virgin Express would disappear from the radar screen.

Question 3: How does the next corporate evolution affect the staff?

In March 2004, the announcement of the discussions with Virgin operated like a tsunami, which massively shook the downward trust-building mechanism. Via the press and internal communication (e-mails), staff were informed that SN and Virgin Express had signed letters of intent to explore the opportunity of a common future. This resounded like a thunderstorm in the small world of SN. At the executive level, the 'Reds', as SN staff called Virgin in contrast to their own 'Blue', were what gave them heartburn. I remember the EVP Network being nearly knocked out by the announcement, with people saying, 'It is not possible, you don't know it, but when SABENA went bankrupt these guys drank champagne . . . it is not serious! They will threaten our turnaround and what will BA say, it's a form of betrayal . . .' It is fair to conclude that the staff felt lost.

THE PRIMARY MISSION IS COMPLETE, AS IS ITS TRUST- AND CONFIDENCE-BUILDING MECHANISM

The board anticipated the achievement of the three points of 2002's mission. Their implementation did not represent a critical issue any longer. The new mission related to equity followed by integration. Today, I consider that the CEO and the Executive Chairman saw their confidence-building efforts shaken to their foundations by the Virgin Express announcement and the length of the negotiation period, because fundamental existential questions were raised again among the staff. The silver lining of this step was that the momentum of the dynamic kept delivering improved performance. But the new context was different: a larger organizational perimeter but managerially speaking a company nearly back in the early days of the relaunch. The staff had emerged from SABENA's ashes, started again in a rather blurry context, clarified it thanks to inspiration (correct relaunch choices) and especially their own efforts, and now they were back in the fog. This created more hurt than disloyalty.

What the SN team was able to do was to keep focusing on its business model of performance improvement, which operated as the best repellent against doubts during the second act. But the confidence-building process would have to resume all over again on a radically different theme and dynamic when a new CEO finally joined the company to write the third act that the board and its President desired.[18]

---

[18] SN's new CEO finally joined the company in September 2006, nearly one year after Rob Kuijpers' dismissal, and was from the telecommunications sector.

BEAR IN MIND

A CEO acts as the hinge of the corporate confidence-building process, especially in a turnaround. This supposes that the CEO can rely on a stable system of management values that he or she has enforced through challenge after challenge. For instance, Peter Davies kept enforcing a system of three intertwined values: truth to build trust, pull to nurture leadership, and empowerment to demand accountability.

Enforcing any value system in a freshly formed organization can be divided into three steps:

1. Share your own value system with the company's senior management to ignite change. To guarantee some lasting impact, more than words is required. Offering access to the intimacy of one's own value-building mechanism is necessary to give not only a clear proof of commitment to the organization, but also a demonstration of closeness with those who are to be led.

2. Prolong the enthusiasm by supporting the disruptive or rejuvenated values through some strong symbols.
3. Repeat, repeat, repeat the message at any opportunity of
   contact with the company's stakeholders. Transforming an
   organization is first about being systematic, not brilliant or
   insightful. The next steps can follow once the foundations
   have been consolidated.

The confidence-building mechanism works in two directions
vis-à-vis the board and the staff, with equal importance. In the
upwards confidence-building mechanism, it is wise to recognize
the reasons for which the CEO has been chosen. Is he or she the
right person or just a necessary evil? The answer to this very
straightforward question represents a form of acid test for recognizing from the start whether both the CEO and the board are
transparently beginning on the same wavelength. In my opinion,
the responsibility for provoking this assessment and obtaining this
answer lies on the new CEO's shoulders, who in this way demonstrates his or her ability to keep a cool head and not be tempted
by the function or the status.

Downward confidence building is a question of context. Is
the appropriate context available to bring the vast majority of
the organization into this survival project without any second
thoughts? Consequently, any form of deviant behaviour or attitude
that can threaten this objective must be relentlessly fought as
a source of waste or even a fatal threat. This demands that those
involved become map-reading experts. The first map helps
qualify the type of behaviour among your own staff so that you
can recognize true fans, and isolate and treat threats of any nature,
while simultaneously investing in the high-calibre but slower
movers who are still needed in the team. The attitudes represent
one map-reading challenge, but the structure is the second one.
Mapping how information channels function quickly helps
the new CEO assess how the downward confidence-building

mechanism can be optimized or penalized, and incidentally where he or she must invest energy to obtain the highest return.

The turnaround mission must not obscure one of the key roles of any CEO: promoting the company's strategic evolution. This must be on the CEO's agenda very early on in order to be sure that the board also keeps reading the priority of the challenges in the same way. Andy Grove reminds us that the CEO must also be a permanent devil's advocate. Finally, the first positive financial results, even though the earlier they come the better, do not mean that the turnaround's success can be taken for granted. The risk of backsliding just increases if the CEO doesn't keep hammering in that one positive financial result does not automatically mean a recurrent series of them, unless a management development dimension is given to the turnaround. In my opinion, promoting strategic evolution, being devil's advocate of one's own role, and communicating actively upwards and downwards is a combination that prevents the CEO from being sent off before the end of the game!

PRACTICAL REFLEXES TO GAUGE YOUR TRUST- AND CONFIDENCE-BUILDING APPROACH

Upwards

- Check with the board the reason for your choice as CEO, so that you both start on the challenge on the same footing and the same wavelength.
- Remain the promoter of or at least an active contributor to the strategic thinking process and its associated decisions.
- Pay attention to when the whistle is blown – the length of the game is influenced by external factors. This is never a game with a predetermined length, therefore it demands even more vigilance.

Downwards

- Cultivate closeness with the management team.
- Map the leverage potential of your closest collaborators.
- Eradicate deviant behaviour and attitudes, which, as CEO, you judge to threaten the transformation process, its deliverables and its schedule.
- Promote your own values continually, and support them with appropriate symbols.
- Surprise people with disruptive behaviours to break routine, even in a turnaround.
- Nurture the transformation process, protect your staff from anything that can divert them from the ultimate goal.
- Promote individual achievements to support the team's cohesion.

COMMENTS AND OBSERVATIONS

Sébastien Bazin, CEO Europe, Colony Capital[19]

For a private equity firm like Colony, permanent confidence is the backbone of our agreement, our contract with each of our respective investment's CEOs to raise the value of the firm he or she will manage within a time frame of four to six years.

In the context of a turnaround, a CEO is, as Jean-Frédéric Mognetti writes, the hinge of a trust-building mechanism connecting the company's board and its staff. But this situation rapidly evolves into a more complex mechanism when the turnaround is operating successfully, or in the case of a well-performing company. Once a turnaround has been achieved, the CEO quickly becomes the critical part of a reconciling mechanism of barely converging expectations between the two stakeholders mentioned above, but also the customers, the financial analysts in the case of listed companies, and even some additional ones such as unions or

---

[19] See www.colonyinc.com/europe.

representatives of public bodies. Consequently, whatever the nature of the challenge, a CEO runs an information clearing house where all the stakeholders must find the reasons for renewing their confidence in the organization through what he or she wants to execute.

'Information-clearing-house manager' may sound an obvious concept, but its sustainable and successful implementation requires avoidance of a dangerous trap. By information, I mean information to feed a dialogue, not just reporting, which I take for granted and does not aim at the same goal. Managerially speaking, dialogue can even be fake. But for me it becomes something distinctive through its systematic occurrence without any complacency about not doing it. This creates the conditions that lead to superior results because the CEO/board team anticipates, then anticipates earlier, and consequently reacts faster and more appropriately. One must never forget that success in business means doing things better than your rivals, particularly in fields with great competitive leverage.

Dialogue does not happen by accident. For private equity investors like us, it raises a crucial question: how much time have we invested to create the right context for interaction? This calls for precision. How much time did we spend – did I spend – listening to the future CEO(s) in order to be mutually convinced that we are on the same wavelength for this investment? Do we read the detailed map of this value enhancement, in this specific context, in the same way? The suggestion made by Jean-Frédéric Mognetti that future CEOs must ask the board which characteristics of their track record and skills made it offer them the job is paramount. I consider this a mandatory acid test that cannot be skipped. Neither party can afford a casting mistake.

This rule concerns the appropriateness of the CEO's profile for the mission, reminding us that creating value is a must. So, execution is critical and constitutes the foundation of continual renewal of the confidence-building mechanism. Based on my experience and some observations, I would like to share an important warning. Colony is not always a majority investor and, in some situations where we are a reference shareholder in a fragmented shareholding context, I too often see new CEOs who believe that their achievement consists in becoming CEO. This is wrong! It is just the beginning of a longer story. Consequently, there is a need for intense dialogue, which leads to mutual agreement for a predetermined period, and for the same intensity of communication in both directions once the CEO gains the arena with a new mission. At this stage, a CEO is accountable to the board for maintaining strict discipline regarding the milestones of the specific roadmap or the business plan. That does not mean that its elements are frozen, but we need to discuss them in relation to the plan. Dialogue also has a formal dimension.

Colony's LBO (leveraged buyout) contexts are intense dialogue builders. In an LBO the debt is a superb repellent to complacency, and operates as a constant reminder that superior-quality execution and consequently results are required. Successful CEOs usually deal with areas of their own resources that have previously been untapped to make the LBO a success. The context shapes the management leverage. It is precisely at this stage that the dialogue with a permanently available and committed board is so crucial for future success. The CEO has broken out of their previous comfort zone as head of a division or department of a multinational company to become the CEO of an LBO operation. The pressure of the challenge is inherent to their new context of execution, so they need to find among the board a substantial part of the fuel of their own dynamic. A board in a private equity context doesn't want to be reported to, they want a continual dialogue.

This can be illustrated by these words from Konosuke Matsushita, the founder of the eponymous group: 'Big things and little things are mine, the rest can be delegated.' We are not here as a substitute. The CEO leads their own staff; that is not Colony's role, nor does it have the resources. Our boards are here to provide the intellectual challenge and to share their practical experiences to help the CEO gauge whether what they intend to do still makes sense for the success of our common project (see the comment on a future-driven board by Magnus Welander, page 289). To play this role of coach, the expertise of the board on the business model as well as our general experience as a 'sounding board' are paramount. This saves time in discussions, it reassures everyone and it allows us to reach the core question faster: how to boost the competitiveness of our investment and consequently its performance.

The frequency and the transparency of these dialogues develop confidence, but this formula is fragile and needs to be continually nurtured. We are greedy for information, but the initiative must always come from the CEO. The first warning signs in this quality of dialogue come from a declining frequency in contact. Then, one starts to imagine reasons for this, which most of the time are ill founded, and it begins to spoil the relationship. The idea of contact refers to the informal dimension of the relationship, where one is able to talk about apparently nothing and coincidentally bump into a real hot topic. Salespeople say that genuine meetings begin on the way back to the elevator. It is the same management style that I support and enforce with real effectiveness.

This continual dialogue between the CEO and the board, the CEO and their internal management, extends to some other stakeholders through a genuine open-door policy as well as a management style of wandering around. It is a huge investment, but it is what is expected from a CEO leading an LBO or a turnaround. At Colony we estimate that this role can consume up to two-thirds of a CEO's time. Nevertheless, building the future also requires free time to think: five to six hours a week, usually alone, when problem-solving skills are left at the door and there is space to rejuvenate one's own creativity.

CEOs of LBOs are solid managers both through training and experience, with obvious charisma expressed through their communication style. Accountable for their actions, led by ambition both for themselves and the company, their greed for success needs to be satisfied by a fair way of sharing results. These creative people know how to question themselves on the appropriateness of a decision, driven by the idea of never making the same mistake twice. Consequently, they have enough humility to keep listening to others. In summary, these men and women are charismatic leaders who have the courage of their convictions.

PART II

THE DRIVERS OF A
TURNAROUND STYLE

*T*he Limelette seminar launched the effort to invent the new corporate tempo, to be between the most punctual and the most profitable airline. The personal accountability of the senior managers was the foundation stone of SN's collective accountability for turning this challenge into a success. Peter as CEO was also conscious that he needed to make his own contribution to the turnaround challenge in order to sustain momentum over the critical period of the first 12 months. In his opinion, the style that SN needed to develop combined four key principles: closeness, building the right environment, vision and speed of execution.

In this part I will illustrate some key steps in implementing the right style for achieving a successful and effective turnaround.

CHAPTER 4

CLOSE AND CONCENTRATED

BUILDING THE RIGHT CONTEXT

History can help remind us that the business context, while it can be very scary, is just one element of the whole picture (see Story 4.1). The decisions that managers make are what influence the execution of a strategy more than the context. In a turnaround execution is even more critical; consequently the manager or CEO must be in the middle of the arena, leading the company, staff and management to make a difference. Turnaround is a topic that fits perfectly with these words from the French philosopher Montaigne: 'To achieve big things you must be among the men' or, in the original, 'Pour faire de grandes choses, il faut être parmi les hommes'.

Story 4.1: El Alamein

On the eve of the battle of El Alamein – which was decisive in the outcome of the African side of World War II, stopping Rommel's troops' progression in Egypt – General Montgomery gathered his commanders in his caravan to explain his plan. There had been a series of defeats under General Alexander's command that had led to many losses in terms of troops, weapons and equipment. Monty's plan sounded impossible to his commanders, who stressed the low morale of the troops after what they had endured.

Monty took out of his pocket a length of cotton thread, stretched it and put it on the table separating him from his group of senior officers. Then he asked one of them to push the thread in his direction, keeping it straight. One of the senior officers took up the challenge, but he could not achieve it. Monty commented, 'You cannot make it work that way for the same reasons as you do not feel comfortable with this plan. This plan, like the cotton thread, demands to be pulled. The troops with low morale need to be pulled not pushed to find the energy to keep fighting in these very adversarial conditions. Your role is to take the lead and to pull them to a victory.'

Any CEO embraces a turnaround challenge with a great ambition for the company, for instance restoring it to a prime position in its sector. But whatever the ambitions are, they have little chance of being achieved if they are not addressed with a great deal of respect for everyone else in the company. Attitude and behaviour show this respect far more quickly than one may imagine. Jean-Marie Descarpentries' first measures as Bull's new CEO (see page 17) not only illustrated the requirements for closeness, but also gave the company's staff a fresh mental representation of how things were going to be done under the new management.

Closeness precedes context building

Those who break with what has gone before also need to stay close to take part in the rebuilding process. That is why Peter ranks closeness as his no. 1 change management leverage tool. 'Closeness is at the heart of my CEO's mission,' he commented. 'It is one of my key themes, which I always develop in any organization to achieve one of my most crucial objectives, creating the right environment for other people to work to the best of their abilities.'

Peter has always described his mission as being like the conductor of an orchestra or, better for him, the coach of a rugby team. I think his mentor was Gareth Edwards, the Welsh rugby captain in the 1980s. The conductor of an orchestra has to manage a team of high-ranking soloists. This perfectly expresses what Peter is avoiding. He tries to get superior performance not thanks to a team composed of great potential talents or even stars, but with a team of 'ordinary' people producing at their peak.

> Achieve superior performance with ordinary people playing at their peak – that is a principle of sustainable success.

Reaching this stage is not an accident. It does not occur spontaneously. A specific context must have been designed. A context is a well-assimilated frame of reference with a strong sense of direction. To support this opinion, I borrow some further words from Andy Grove: 'There is no gain in being able to recruit great employees, handle a board, dazzle Wall Street or rally your cavalry for a glorious charge at dawn's early light if you haven't figured out which direction to point the horses.'[20] Alan

[20] Richard S. Tedlow (2005) 'The education of Andy Grove', *Fortune*, 12 December.

Lafley, the current CEO of P&G, shared his beliefs about the dynamic associated with context in these terms: 'The more you understand something, the more you are willing to take risks and the more intelligent are these risks, then we can stretch ourselves to go for peak performance without breaking down.'[21]

In a turnaround, even more than in other business situations, the prerequisite of building a context is one of the hurdles that absolutely must be jumped by those who want to bring the company back to the correct track. And objectively, Peter Davies was developing this requirement as a form of permanent obsession in all the turnaround or radical business performance improvement situations he was involved in.

Who is the context for?

The turnaround task is not easy, but SN was not Peter's first attempt. His whole career has been dedicated to this kind of management demand, with varying levels of difficulty, from one corporate logistical challenge to another. He has operated in radically different context-building situations. His first experience with XP Express Parcel System (KLM) in the UK could already be called a turnaround, then again with DHL in the UK, then pioneering, fixing and boosting DHL Americas, and in his last challenge with SN Brussels Airlines, which might more properly be named rewiring. But each time he speaks about these challenges, his words about context building are people-centric. As an example:

> My passion to offer to our collaborators the right context was developed along with my career in the service industry. I am a network man who has always dealt with highly perishable consumer

---

[21] Rajat Gupta (2005) 'Leading change', *McKinsey Quarterly*, July.

goods. But, I believe that the nature of the business influences the way the creation of the right context is done. Usually, what has made the traditional workforce productive is the system itself. In fact, whether it was Taylor's 'one best way' or Masaaki Imai's Kaizen, the system embodied the knowledge. Consequently, a highly skilled person in this context can become a threat to the system. But in a knowledge-based organization, like our service industry, it is the individual worker's productivity which makes the entire system successful or not. To summarize, in a knowledge workforce, the system must serve the worker. That is why both by conviction and experience people are central in my management philosophy.

Due to his experience, Peter's management approach is appropriately influenced by the importance of people in achieving a challenge. But in a turnaround context, creating the right environment encompasses an additional constraint. The goal is to make change occur, which means achieving different results; the difference is that in the turnaround context this usually has to be done with the same people, but with fewer of them.

First, a quick reminder of the pitfalls is necessary. A massive substitution of people is never a guarantee of recovery in any business. In many successful turnarounds two phenomena can be observed: first a downsizing and secondly an injection of new talent. The reason is simple. The company must deeply revisit its fundamentals in terms of skills and staff.[22] The diagnosis is very clear: the company does not have the correct numbers nor the appropriate talent to deliver the operating performance imposed by the sector at the appropriate pace. You cannot hide from the numbers. 'We have a structural problem, we have a headcount of 200 per aircraft – American Airlines has 120 and Southwest certainly 90,' said the new CEO of Caribbean airline BWIA to his

---

[22] Staff and skills are two elements of the 7S model proposed by Peters and Waterman: Thomas J. Peters and Robert H. Waterman (1982) *In Search of Excellence*, New York: Harper and Row.

senior managers who were gathered for the first time at a convention. The harshness of this comparison could be read on the
audience's faces. But the real problem is that the numbers have
not changed overnight, they have already been here for a pretty
long time and the only possible way to get back in the game is
to beat them. Slow evolution or expansion is not a feasible option.
In this context radical measures are mandatory or one falls into a
form of guilty complacency.

Swedish company Envirotainer (see comments on Part VI by
Magnus Welander, CEO of Envirotainer) manufactures and rents
out refrigerated containers for airline shipments, mainly specializing in the pharmaceutical sector. At the end of 2005, after two
years of effort, the young CEO, Magnus Welander, posted his first
positive cash flow. This occurred after a massive injection of cash
from investors, nearly 50 per cent of turnover in a company
achieving sales of €20–25 million. Out of a total staff of 165, 100
left and 20 new collaborators joined the company. These figures
describe the magnitude of the challenge when in-depth corrections have to be made to secure a sustainable future for a company.
A last detail is important. This company is the world leader in a
niche with almost no challengers at all. The correction measures
reinforce the idea of who is any organization's worst enemy or
competitor – in the vast majority of cases, it is the organization
itself.

A context to transform losers into winners

In SN's case there was a substantial silver lining, because there was
no requirement for the company to shed a lot of people immediately. This is one of the big differences between a turnaround
before a bankruptcy, where difficult decisions have to be made to
ensure that the company survives, and a relaunch like in the SN
and Swiss International cases. A 'start-through' like this has to be

achieved with the same people in a massively downsized organization; only a sixth of SABENA's staff remained, for example. It is these 'survivors' who have first to be transformed into players before making them winners.

Outside observers may quickly conclude that the mission only had to be completed, the dirty part of it – getting rid of staff – having already been achieved. This was only partly true. Quick and effective action had been taken, with speed and a good dose of effectiveness, but no further structural adjustments were possible for a long time, as in any turnaround. The new company's configuration must fit with the sector benchmarks, meeting the correct numbers with the appropriate quality – that was the challenge.

The relationship between the CEO and the employees was already established and the membership of senior management was virtually fixed, so it was the CEO's people management skills that could make the difference in creating the new appropriate context. But this right environment is just a framework, which turns into a waste of energy unless one can create a dynamic. Building the right environment demands dialogue, not simply gossip.

Context requires dialogue

Discussions can keep going on and on. In contrast, a dialogue is an action-oriented exchange. It does not find its justification in itself. A dialogue is not conducted to spend time, it is intended to make progress. Fairness and transparency are the characteristics of a fruitful dialogue. If they do not exist, no progress can be made. So the goal is to create the conditions for a constructive dialogue, which is characterized by its theme. If there is no vision to share, there can be no constructive dialogue.

Nevertheless, there is a flipside to the word 'vision'. It is a two-edged sword that can boost or hold back progress. Remember the warning from Lou Gerstner, former IBM CEO: 'A vision can be a distraction, so I have never offered one.' Gerstner did not take the risk of having a vision in order not to create more confusion in his organization, while Peter Davies needed to lessen the confusion and show his staff that the direction he was pointing the company in was marked with enough signposts to allow the team to move ahead. The CEO needs to select the right tool from his toolbox like a magician from his box of tricks.

First, create the right context for people to work to the best of their abilities. Secondly, develop the appropriate dialogue with and among senior management and staff.

The challenge can be seen as a question of inspiration and creativity, but these skills are taken for granted at this level of management. In a turnaround, inspiration is not the critical resource, time is. Therefore, a sense of priorities is a critical skill for delivering the right tempo to sustain the momentum in the right context for people to work to the best of their abilities.

DON'T WASTE LIMITED CORPORATE ENERGY: CONCENTRATE ON A SUITABLE CHALLENGE

Don't trade off tempo for bells and whistles. The organization always needs the right tempo to sustain the momentum in its recovery or progress. In SN Brussels Airlines, the staff were traumatized. The company remained in the same sector after what was a veritable industry earthquake, the bankruptcy of two highly visible European national carriers. Consequently, these people didn't dare express their impatience to demonstrate that what happened was neither fair nor a reflection of their real value. So it is the responsibility of the CEO to articulate the correct tempo to rebuild the organization as fast as possible, and boost the staff's confidence in their own future by offering them the correct opportunity to demonstrate it.

Too early for a corporate-wide mobilizing theme

Like sportspeople whose pride has been wounded by a defeat, the collaborators in a turnaround expect to get a quick opportunity to improve the odds in their favour. One of the first checkpoints on the CEO's list is immediately to assess whether a large majority of the team can be quickly remotivated. In such a context the size of the group to be remotivated is a critical issue: one needs to know if one can quickly count on a large enough group of fans to succeed in instilling the new management style and approach. The challenge is not only to choose the correct treatment but also the appropriate dose. At SN, the critical size of the group was easily reached due to support from the management. Therefore, once the group was defined, the next objective was to bring it together through a mobilizing theme. This helps focus the energy of the transformation group and at the same time helps its members recover some confidence in themselves and consequently in the company.

Peter and I discussed the issue of a mobilizing theme often, and found an analogy with a spectacular turnaround in the float glass industry in Liège at a company called Glaverbel. The mobilizing theme that Jean-Marie Descarpentries invented in the early 1980s was quite chauvinistic: 'beat the Germans'. The reason was simple. A German company just on the other side of the border was the most productive float glass manufacturer in Europe, and Glaverbel was among the top low-performing companies. This theme, which the whole company including the unions readily accepted, addressed the deep pride of the team in order to exploit untapped energy that had been wasted by years of internal conflict. The leveraging mechanism was not new but remains very relevant. So my recommendation is not to waste energy rectifying internal disorganization, but to test whether a disruptive theme focusing attention on a radically different objective is able to focus energy and generate initial recovery success.

> **A mobilizing theme is a focusing technique to optimize limited energy.**

It is not always possible to find a mobilizing theme as simple as that mentioned above. The risk in the airline sector was to have too generic a theme. In *Organic Growth* I stress that executive committee members must be involved in developing growth opportunities by searching for a weak signal in the sector. So instead of brainstorming resulting in potentially mediocre themes, do something disruptive.

There is a risk in a copy-cat approach without assessing the relevance of the managerial tool being used to the company's current context. At the corporate level at this stage of SN's evolution, it was not appropriate to use a mobilizing theme, but that does not mean that this concept did not have to remain at the forefront of the CEO's mind. A few months down the road, when we formalized the strategic dialogue tool at the elementary level of strategic execution (see Part V) – 'the route' – the mobilizing theme was introduced as a mandatory step of the diagnosis and action-planning process (see Figure 11.3 on p 206). It helped focus the energy of the small group of committed people who were accountable for the route's performance. When managing this kind of tool, one must always pay attention to whether the level of mobilization one intends to reach is appropriate in terms of both context and scale.

Finally, at SN the situation was more complex due to limited corporate energy. One runs the risk of wasting energy if the theme is not virtually 100 per cent correct. So I recommended not using this tool of a mobilizing theme at this early stage. It was wiser to try developing an alternative proposition to capitalize on the real thirst for action in the team, by involving senior management in something that the company had never done before.

'Never been done before' fitted the context better

A new CEO – and Peter was no exception – continually receives messages from their senior executives about their ambitions for a fast recovery or improvement, with suggestions about new services or new routes to develop. This is a classic illustration of the confusion between speed and haste. The key challenge in this case lay elsewhere. Did SN have a common and shared view of the reality of the relaunched business challenge? The page had been turned on SABENA, and this was not because the company name had changed and a few comments had been made in the press. Providing senior management with the possibility of getting to the reality of this situation is attractive, because it is a means of doing things differently and being consistent when one exhorts the company to be transparent. This starts by the CEO being transparent and daring to share this lucidity.

The choice: a different kind of customer survey

My idea was to funnel the thirst for action of SN's 40 senior managers into something that would help improve the company's strategic knowledge. During the Limelette seminar – which took place six months after the relaunch, four months after Peter's arrival, with the first ramp-up phase already behind them and some real success – the goal was to translate the discussions and conclusions on the theme 'Inventing the tempo between the most punctual and most profitable' into practical terms. Everybody in management accepted the correlation between punctuality and profitability. But such a discussion is academic if one has not understood that as a manager, one is both the cause of and the reason for any success. The behaviours and attitudes of one's own department are part of the final result and the only way to reinforce this result is to invite the customer into that department.

But in such a context, this step was many bridges too far.

As a warm-up exercise, the suggestion was to go and meet customers to discuss the new corporate characteristic of punctuality and to explain that their answers would be critical in supporting the momentum to deliver the company's promise to its customers.

My suggestion was pretty basic: a customer survey. Some observers might have wondered why this had never been done before. While SABENA had certainly carried out hundreds of customer surveys, SN Brussels Airlines had not yet done so. Our ambition was always to find innovative ways for doing something that already seemed very familiar. 'Never been done before' (NBDB) finds a lot of its strength in rejuvenating good old and familiar practices. So the question was not to know that such an approach had never been done before, but whether the 'corporate doers' had done it before. For this survey, the internal target was the senior management team. Therefore the difference lay in the conditions of execution, in those doing the survey. This was not a management gimmick, it was critical to instil a new sense of corporate contribution in a synthetic measurement of organizational effectiveness.

> **Never been done before by the corporate doers.**

This senior management customer survey had many advantages:

- It prolonged the discussion at the conference about the company's new market legitimacy and protecting a coveted result: 'most punctual'. This result meant something to the CEO and senior management. It represented a fact that reinforced the direction in which the company had chosen to move.

- It was a practical and cheap tool to close the gap between company and customers from an angle that had never been used before: a survey conducted by a team of two senior managers from different departments.
- It sent a distinctive message to the business community about the effectiveness of the approach of this new company that was still struggling for survival.
- It also represented an additional opportunity to reinforce the feeling of belonging to the same team looking at the future with a common shared vision. The managers involved did not have daily contact with each other, so the survey helped break the silos in which people continued to operate.

This approach must not be read as a means of making the company more customer centric. That would not have been realistic. At this stage, such a generic management recipe could have caused more organizational damage than advantages if it was not appropriately managed.

Senior management in the field

So in teams of two the senior management went to visit three targeted customers in the vicinity. Many untapped resources almost always exist near any company, which offers a valuable management development opportunity with limited logistical constraints. SN Brussels Airlines is located near the airport of Zaventem on the outskirts of Diegem. These two communities represent one of the highest European concentrations of headquarters or coordination centres (global or European). Many of these companies are SN Brussels Airlines' clients. So the goal was to organize a meeting with those responsible for the travel policies of these companies. This involved contacts ranging from senior purchasing managers to CEOs.

> Getting out is everything to becoming more forward facing and gaining in business lucidity.

But there was something unique or least intriguing for the targeted customers of the survey: they did not receive a call from the salesforce or from some specialist consultant in this kind of activity, but from a back-office or operational senior manager, including SN's chief pilots. 'I am Bart, chief pilot at SN Brussels Airlines, and I would like to have an interview with you to discuss your company's travel policy.' This was something customers admitted they had not heard before. Some of them even said to SN's executives that their organization should do the same thing. The key point of the discussion over the phone was to explain why the survey was being conducted by operational rather than sales staff. Once a simple explanation about punctuality had been given, the appointment was set up without any difficulty – there is always some curiosity to benefit from – and 60 interviews were conducted over five weeks by the 20 teams. A debriefing session was planned for early December.

Easier said than done!

But this kind of mobilizing process does not occur just because the CEO decides it will. The management task of sending 40 executives in the field can be divided into four hurdles:

1. Overcome the fear of some executives. A few of them told me that they were not ready to do this exercise because they considered that they were not the most appropriate person. Faced with such a corporate challenge there can be no exemptions – it is for the whole team. The preparatory steps will diminish the fear. These people left the company shuffling their feet and came back enthusiastic.

2. Benefit from team building. Building 20 groups of two people was important not only in interpersonal terms, but also to pull down walls in the organization. I noted very early on that even though all the managers came from SABENA, not only did they not know one another very well, in the vast majority of cases they did not know each other at all.

3. Develop a detailed map to structure the preparation, the interviews and the debriefing. This document was important for running the briefing with each team, and reinforced their confidence that they could achieve the task professionally. But I also anticipated that it could be a form of marketing exercise if the customer asked to see or read it, which did happen.

4. Do not underestimate the importance of the debriefing step. A debriefing was done with each team. But in order not to interfere with day-to-day priorities, four or five debriefing slides were prepared by my team of MBA students and myself. Once the three interviews had been conducted, in the executives' mind their mission was over. They did not see it as their responsibility to make the transformation happen. So in order for something to happen – as we will see very often with other tools developed during this project – it was the responsibility of the CEO and his restructuring team to provide the tools and the support for a new type of management game. With the collaboration of my MBA students from HEC, we controlled the style and the content of the last but one step in the exercise: the senior management debriefing, which took place eight weeks after the decision to launch the survey. The purpose of the small debriefing presentation was twofold:

   • To provide support for a discussion with senior management colleagues and the CEO.
   • To gather all the members of the managers' respective departments together to share and discuss how customers see the young company and what had to be done organi-

zationally to keep enjoying the coveted result of being the most punctual company.

A fruitful crop

This survey was conducted just a year after the bankruptcy of SABENA. It was an occasion for assessing how much sympathy there was for the new company, but simultaneously for noting how fast that sympathy could also be eroded. Moreover, it brought an answer to a crucial question: who is the real enemy? One answer is obvious: SN Brussels Airlines itself through its own behaviour. But who did customers see as the benchmark?

> **Who is the enemy?**

The answer to this question was very important to break a preconceived idea. SABENA was dead, but the internal reflexes of a former hub operating carrier were still present. Part of senior management, the network for instance, considered that the 'enemy' consisted in the other traditional carriers such as the members of STAR Alliance, for example Lufthansa or SAS. But from the survey, the dominant alternative option was Virgin Express.

This result was very interesting, because the market's perception of who was SN's enemy was in fact wrong. The market was being influenced by the low-cost fad sweeping through Europe at that time. Virgin, with its aggressive and effective communication strategy, easily became top of mind in Belgium. But the company's performance was mediocre. For instance, it suffered a financial loss in 2003 when SN cut its block seats agreements with the company. The STAR Alliance was the real enemy and could have become even fiercer with its purchase of Swiss, SN's historical code share partner.

From a management point of view, the conclusions of this kind of survey must be regarded as raw material for further internal discussions, not as a crystal-clear strategic diagnosis. After such an exercise, it was easier to start mapping out the company's real competitive challenge: how much sympathy capital SN enjoyed, and what had to be done in order not to waste it. For instance, this practical form of market contact between customers and SN's 'doers' was extended to the cabin crew, who on a volunteer basis joined the sales reps to perform joint calls on travel agencies in order to develop a different image of SN among its distributors.

> **Pushing broader and further market contacts.**

Proud of the results of this first attempt, it was immediately suggested that the NBDB approach should become a 'best demonstrated corporate practice'. We took the decision to organize a survey on a biannual basis, with a specific target for each run. The next survey in Spring 2003 focused on the customers of SN Brussels Airlines' outstations.

> **A new best demonstrated practice is born.**

The objective of this approach was to alleviate as soon as possible part of the deep wound of the past trauma. But at the same time, the company's resources and energy were very limited. So don't break the bounce of your organization by mobilizing it too early towards a difficult goal, but redirect its limited energy in the correct direction through a 'cluster breaker' exercise. The CEO needs to create a context where the efforts of the senior management team have a clear impact in terms of confidence building. Finally, this will also reinforce the corporate feeling that in the new context a different tempo in the relationship between senior management is spreading all over the company: 'Who did you do the survey with? Who did you meet? Did they tell you that . . .?'

This kind of corporate practice is not effective just because it is publicised in the management press. Before thinking in terms of benefits, it is wiser to think in terms of turmoil. There is nothing obvious in changing one's own role to become a character in a scenario that one has not written. This approach was a practical way to force people to think out of the box. But it is ridiculous to think that it was obvious for a team of 40 talented managers in their respective areas. The hurdle of the inhibitions to be overcome must not be underestimated.

So the recommendation at this early stage of a turnaround is: don't waste energy in formalizing a sophisticated vision, instead involve senior management in practical actions that represent a disruption from their past history. This right context becomes a commonly shared platform, which can reap the dividends of practical actions in the sense of the turnaround.

Not promoting the corporate vision is not a way to burn some sacred cows, but a way to prepare the team for excellence in execution, which is a far more demanding task than just building a vision. Lawrence Bossidy, the retired Chairman of Honeywell and former CEO of Allied Signal, remarked in an interview with *Fortune* editor Geoffrey Colvin on the magnitude of the execution challenge:

> You can exhort all what you want about excellence in execution, you are not going to get it, unless you have disciplined strategic choices, a structure that supports the strategy, systems that enable a large organization to work and execute together, a winning culture and leadership that's inspirational. If you have all that you'll get excellent execution.

CHAPTER 5

PRACTISE EVERY DAY

DAILY EXERCISES TO CONSOLIDATE THE RIGHT CONTEXT BY MICRO-MANAGEMENT

The right context is a critical step to enable the organization to start on the correct foot. This context is supported by the fresh sense of direction provided by the new CEO, combined with a 'never been done before' exercise working at the macro level of the turnaround. At the micro-management level, a different set of attitudes and behaviours also has to be enforced to reinforce the new direction until the first positive financial results have been obtained.

SPEED OF EXECUTION WITH A SENSE OF URGENCY

Don't count on having time to observe! Speed of execution is crucial. You need a plan from the beginning that makes the new priorities clear.

Jean-Marie Descarpentries made this interesting comment about speed of execution six months after he took the helm at Bull, speaking on my MBA program when I was dean of the European School of Management in Paris: 'It took me 100 days to achieve a diagnosis and deliver an action plan in Carnaud. In Bull, 15 years later, nearly four times less, 28 days.' He added, 'The reason is simple, for a very limited part it is due to seniority, but for the most substantial part it is due to experience . . .' He stopped and one MBA student finished his sentence, saying, '. . . experience at delivering a diagnosis.' Descarpentries looked at him, smiled and added, 'Experience marginally helps in the analysis, but it gives you an even more sensible sense of speed, or urgency in this critical period.'

> **100 days at Carnaud to launch the turnaround plan, 28 days at Bull prioritizing a sense of urgency – that explains the difference.**

A turnaround is like an emergency operation where one is working against the clock. The CEO is in the role of surgeon. The organization is traumatized, just like an accident victim. Consequently, the organization needs to be treated with tact, respect and empathy.

The analogy with an emergency physician allows us to perceive the nature of the first reaction one has to have in such circumstances. Doing nothing is not possible, and increasing the chances of the organization surviving also relies on the first professional gesture from the CEO. While empathy is key, the necessary management style cannot be achieved without a substantial dose of autocracy, legitimized by a combination of one's own demonstrated skills and accumulated experience. Some emergency measures must be taken to stem the losses – to stop the haemorrhaging – but in the decision to cut one must not confuse permanent damage with injuries that can be repaired. Early and effective downsizing must be done with insight.

'Paralysis by analysis' is a well-known trap. But intuitive managers are also threatened by relying too much on their intuition, which is so often summarized in the famous expression 'Ready . . . fire . . . aim'. To be realistic, intuition must be supported by a combination of hard facts, accumulated experience and speed of analysis. This was perfectly summarized by the French biologist Louis Pasteur, when he said, 'Chance favours the prepared mind.' How and when do you reach this level of preparation? It is in fact this level of preparation that leads to the perception of intuition, because the formal steps of the analysis have disappeared. Therefore the CEO can say, like Winston Churchill, 'My best improvisations are those I rehearsed the longest.'

Intuition combined with hard facts allows one to have the courage of one's convictions and to avoid groupthink, where everybody is following the dominant 'doctrine'. This is fighting against overconfidence, a herd mentality and false consensus. A solid analytical approach allows a sound dialogue, while intuition alone leads to a never-ending passionate discussion, which does not allow the development of a learning organization. Genuine leaders are never afraid of analysis that contradicts their intuition. A constructive Cassandra is their insurance, which allows them 'to keep staying one decision away from disaster', as Andy Grove put it.

MENTAL REPRESENTATION EASES THE TRANSFORMATION PROCESS

Even when there is an overwhelming need for the organization to change, the CEO will still meet with a substantial dose of resistance, because the unknown is frightening and blocks change.

Above, I stressed the importance of the right context as the backbone of a corporate dialogue. Its purpose is to provide milestones towards a new comfort zone. But following these milestones is not something that the whole organization can achieve spontaneously.

The first moves will naturally encounter a substantial level of resistance due to a paradox: the organization knows that its previous mental representation is ineffective and inappropriate, but without a suitable substitute, it sticks to the previous model even though it failed. Professor Robert Ballon, a Belgian Jesuit Father from Sophia University in Tokyo who has been living in Japan since the end of World War II, provides me with a very relevant observation, which suggests that mental representation is already the heart of action. His image is drawn from Samurai legend, where the warrior is supposed not to know fear. In the Samurai doctrine, not knowing fear is just a consequence: the samurai has internalized or mentally represented the fight so that almost nothing can happen that he has not imagined before and for which he has not produced an appropriate answer. To translate this idea into our current business context, it is what entrepreneurs do through their strategic forecasting exercise[23] that develops clarity in assessing the nature of the challenge.

What is frightening? The unknown. Consequently, in this turnaround period the CEO is the pathfinder who leads the company and must take care not unwittingly to create situations that will inhibit change. Consequently in his or her messages the CEO embodies what will transform losers into players and how that will occur. Being winners is the next step, which might occur and to which the CEO is committed. But one must be realistic and not deceive oneself with words. The change process begins

---

[23] A strategic forecasting exercise will be addressed in Part VI, showing that it is neither budgeting nor strategic planning.

when what the CEO is saying can be translated into a mental representation of the challenge for each individual. Then the challenge becomes accessible.

LOOKING FOR THE SILVER LINING: A DISRUPTIVE MANAGEMENT ATTITUDE

In a turnaround context one always inherits something pretty dark, such as the trauma of bankruptcy, but there is also a silver lining, which the talent of the CEO can leverage. In all Descarpentries' turnaround project communications there was always a section called 'Good News'. Experience tells me that it is easy to miss this powerful leverage point.

Its meaning is very simple. From the beginning, one sets up the recovery to evolve along the lines of that silver lining. Secondly, it is an easy and cheap means to show one's own optimism, counting on the fact that good news spreads fast. Finally, one must be selective with what is regarded as good news. Employees are very sharp on this kind of assessment, so the CEO must ensure that good news is always expressed within the terms of its impact on the turnaround. To sustain the momentum, the CEO must make clear that what counts is not the extent of the good news but its impact on the change process.

The CEO's accumulated experience and skills are part of the reason for success. But he or she has an ally, whose impact must be nurtured and leveraged: luck. That is why it is always critical to have a clear reading of the situation, and continually to consider whether what seems a handicap is actually hiding a superb silver lining. That is not only a mindset but a vital attitude in circumstances where one needs to develop a talent at producing something from nothing. Once again, the dividing line between the best and the worst is very fine. The genuine value of the CEO lies in making a difference in this turmoil.

It is difficult to imagine anything darker than SN in 2002: in the aftermath of the 9–11 terrorist attack, with 'no-frills' airlines the new idols of the market. The press was already announcing the end of traditional carriers such as Air France or British Airways in Europe. They did not count on the legacy carriers' ability to fight back in any aspect of their business, from product offer (price) to legal fights over unfair competitive practices (subsidies from local airports coming in the form of a joint marketing budget, for instance). Or take Richard Branson, who claimed that the person who had the biggest impact on him was Freddie Laker with a recommendation summarized in only three words, 'Sue the bastards!'[24] An effective enemy can also learn from your attitude. The rationalization efforts of the major Europe-based global carriers (unit cost reduction, simplification of their own low-cost offering, withdrawing from some 'must fly' routes), combined with a rejuvenated competitive fierceness, showed the sector that they were premier league players capable of adapting. This was not the case for smaller players such as Alitalia or Austrian Airlines, as Professor Doganis illustrated in *The Airline Business*.

In this environment, SN Brussels Airlines nearly disappeared from the sector's radar screens. Retrospectively, its bankruptcy was in fact the lucky event that allowed it to undertake a period of intensive care in relative peace. In the early days before Peter arrived, some good strategic decisions had already been made or were about to be, and this was a critical factor.

Macro economy

The macro-economic conditions were good. The exchange rate of the euro against the dollar was favourable and the oil price was below $30 a barrel. It is quite certain that if the oil price had been

[24] 'The best advice I ever got', *Fortune*, 21 March 2005.

at 2006 levels and the dollar/euro exchange rate below 1.2 SN
Brussels would not have been launched, because the macro-
economic environment would have represented an excuse for not
doing so. This would have been quite legitimate, because the
increase in the oil price would have immediately erased part of
the savings due to the bankruptcy, which allowed unit costs to
fall by nearly 15 per cent compared to SABENA. For instance,
year on year 2004 to 2005, at the end of the first quarter the
average direct operating costs per flight in Europe had increased
by 12 per cent.

Africa

This area was a legitimate and established service for SABENA
that was positively perceived by a loyal customer base. Therefore
leaving it under the exclusive service of another European major
carrier such as Air France would have represented a suboptimization of opportunity.

Code share

SN needed partners to avoid spoiling its potential for success. The
relaunch had been planned as a code share, with Swiss for instance
on the three Swiss destinations. SN planes flew for both partners
exclusively to Geneva, while Swiss reciprocated with its own
planes on the routes to Basel and Zurich. This code share situation
needed to be handled with care. SN also relaunched its operations
to Rome, Barcelona and Madrid with a very costly 'hard'[25] block

---

[25] A hard block seats agreement means that on each flight a set proportion
of the seats is bought by a code share partner, which takes all the risk
to sell these seats. All the other forms of partnership tend to mitigate
this risk with the plane's operator.

seats code share agreement with Virgin. That agreement ended in March 2003. This agreement was a great sleight of hand in negotiation by Virgin and was in fact an effective means to subsidize Virgin's operating performance on that route. Happily, this was stopped as of the end of March 2003.

Fleet

The fleet perfectly illustrates the need to assess the company's heritage. Were there any hidden jewels?

To achieve the new company's mission assigned by the chairman of the board, 'Connecting Brussels', SN Brussels Airlines was relying on DAT's fleet. DAT was the regional carrier of SABENA. It was not caught up in the bankruptcy and represented the relaunch platform for the new company. The same approach was followed by Swiss, which was relaunched from Swissair's regional carrier Crossair, which also had stayed *in bonis* on the Swissair side of this disaster. The European fleet was DAT's, made up of 32 Avro regional jets with a capacity of either 85 or 98 seats.

When I started collaborating with SN, the comments on the fleet were not good. Being a generalist, I didn't bother at this stage of the relationship to challenge such a commonly accepted opinion. Andy Grove has a very sharp formulation to characterize this situation: 'Without any data, an idea is just a story – a representation of the reality and thus subject to distortion.' To become familiar with a new organization quickly one must read a lot on top of discussing with the broadest possible range of people. One day, I read the third of the management plans that were issued every six months by Erik, the EVP Network. I was surprised to find the information outlined in Figure 5.1.

The capacity of the Avro (85 to 98 seats) suited Brussels Airport's average number of passengers per flight. This was very good news: the asset base was structurally perfectly suited to match the

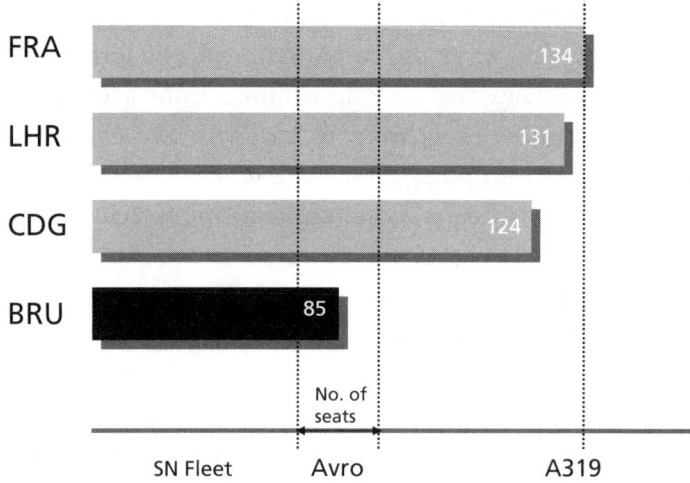

Source: Brussels International Airport Company, 2002.

Figure 5.1 Average number of passengers per flight, Frankfurt, London, Paris, Brussels.

ambition of regaining traffic in Europe from and to Brussels. Was this something new? Was this report the first one? I researched the previous report and I read exactly the same information. So this company was strategically healthier than it thought. But I was astonished that this jewel, although it was known, had not been used to its full impact before, and had even been ignored to some extent. This is one of the strengths of a generalist outsider: he or she can reprioritize some pieces of information that are buried under layers of routine or sector perceptions or preconceived ideas. Stressing its importance before would have allowed the CEO to say formally and earlier to the staff that the company was appropriately equipped for its European challenge. The company had missed the opportunity to exploit legitimate good news strategically.

The organization itself can also become a potential destroyer of good news. A colleague of mine was involved in the turn-around of the Moulin Rouge, a famous cabaret club in Paris.

Surprise, surprise: this kind of place is driven by the same rules as any business. It was close to bankruptcy. His turnaround scheme was based on cutting costs by shifting from a dinner-show and champagne-show programme to one with two champagne shows. But he was confronted with another key issue in changing the show, because there was a diminishing number of customers and those who did come were bored with the show, especially the Japanese clientele. After some quick research, it appeared that a Japanese tourist returns to the Moulin Rouge only once every ten years. The true reason that the show was perceived to be poor value was that the staff of the Moulin Rouge wanted something different, something new – not the market. The conclusion was that the show certainly did have to be changed, but not with the urgency expressed by the corporate gossips.

The turnaround of the European operations of SN had not occurred by accident. This fleet currently provides competitive cost conditions. This performance is the result of its age, which has been optimized in terms of financial engineering. Even though the Avro is presented by its detractors as not the best-optimized aircraft in terms of fuel efficiency and cost of maintenance, those who do own it are currently reconfirming their interest in this type of asset by expanding their ownership, while its manufacturer BAe has simultaneously announced that it will also extend its technical support for another ten years. This situation is tactically very important when one is in the middle of a turnaround, because it releases the pressure for the mandatory trade-off decision to be made between renewal and modernization.

However, some other fleet decisions were less consistent. As the new CEO was arriving, the new company was falling into the trap of incorporating mid-range aircraft.

At the time the press was full of announcements that Airbus Boeing was signing the highest ever deal for this type of aircraft. As of the end of the second quarter 2002, the question of adding mid-range aircraft to operate either with higher capacity and/or farther destinations was on the table. This was not a turnaround decision, but already a strategic one. The question was the object of a tough argument between Sales, which said the company need 70-seaters, and Network, which had not yet achieved its own revolution and was defending SABENA's legacy in terms of mid-range network. With the Airbus 319, SN could comfortably reach Tel Aviv, Istanbul, Casablanca, Moscow and Helsinki, just like the competition. But reaching a destination is one question, making a profit out of it is another. Even though much of this market for capacity reasons operated according to code share, SN was very far from break-even in low season and just in the vicinity in high season.

This decision to add the Airbus 319 created a bridge between the African and European networks' pilot populations. Obviously, some synergies in terms of resource optimization could be imagined and implemented among the Airbus pilots, while the whole cabin crew was, as of April 2002, flying on both networks. This form of silver lining was rather thin in my opinion, and this sounded more like a means of adding a bullet point on an assessment sheet than a genuine argument with intrinsic weight on the right decision.

The conditions of the relaunch forced SN to retain its humility and to keep a clear idea of what could and could not become a leverage tool during this critical phase. We reported some possible ways to handle such a management situation, but there is something that is true whatever way is chosen: it must be achieved by singing off the same hymn sheet.

Compared to SABENA's former relative Swiss, SN's relaunch conditions were tighter. The limited amount of equity, €150 million compared with nearly €2 billion, forced it to be creative but above all to be pragmatic. It was what we called going back to basics.

This situation forced SN into a management approach that was a permanent lesson of what Peter likes to call the advanced management programme of the 'Micawber Graduate School of Management'. Charles Dickens pictured the character of Mr Micawber in *David Coppperfield* and gave him the words to explain the difference between happiness and unhappiness:

> Annual income twenty pounds, annual expenditure nineteen nineteen six, result happiness. Annual income twenty pounds, annual expenditure twenty pounds ought and six, result misery.

So the lesson was paramount for SN too: cash is king, even in a start-through. The basis of balance sheet engineering is tactical, while cash is strategic. In this context examples can be powerful. I could hardly believe it when SN's Switzerland country manager reported me that in GVA when Swiss was relaunched, since the budget had not been completely spent, it was then decided to renew the office furniture. Cash cannot be spent twice, and when you find an opportunity to save some of it don't disregard it. This is a sound management reflex, which too often tends not to be a standard one.

> **Cash is king even more in a relaunch.**

NO MIRACLE, BUT PREVENT PANIC

The arrival of any new CEO leads to a mutual observation period. By those who hire them, a new CEO is nearly always pictured with characteristics that give them the attributes of a white knight. White or not, knight or not is not the question: it is, how can you manage people's expectations?

The company's problems were not created yesterday, so there is always the hope of turning the page. A form of magic is expected or a kind of miracle. Unfortunately, this hope is misleading, because it means people are thinking that the CEO can miraculously get back to the organization's previous comfort zone

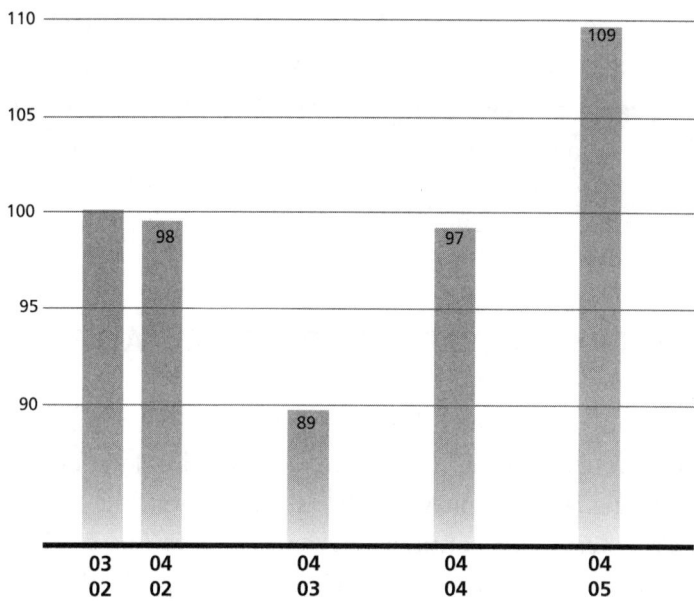

Figure 5.2 SN's average monthly cash position (base 100 launch).

in some ways that haven't yet been thought of. Paradoxically, it is difficult to expect anything else, because the mental representation of corporate strategic health has not yet been rejuvenated.

While trying to organize the turnaround plan, a positive belief in a flourishing future has to remain indestructible. This demands that a CEO and their close team keep a cool head when managing the gap between where the company is already through its projections and where it is now. For instance, in SN Brussels Airlines during the first quarter of 2003, the cash position was moving south, but simultaneously the most effective route management measures had already been implemented. This is illustrated in the curve of the cash position in Figure 5.2.

Consequently, on the one hand building a confidence-development system must be encouraged, and on the other hand the confirmation that the bottom of the pool is close must be permanently stressed in order to know exactly when one starts managing corporate morale by cash. That is why in this context I keep repeating the impossibility of saying whether a CEO in this con-

> **Manage corporate morale by cash!**

text needs to be a right- or left-brained person. In my opinion he or she needs to be both and to hold the helm solidly.

BURY THE GHOSTS, TURN THE PAGE

Turnarounds are never achieved by repeating the past, nor by limiting the exercise to cost cutting. A radically deep break from the previous failed business model is mandatory, but turning the page also means not continually referring to the past. This concerns the management team as well as the troops. In this context, the impact of management by example or the use of symbols is critical. To explain the necessary new attitude, Peter used to say it was easier for him to embody it: 'As a pilot, I have an advantage, I have been trained to look forward.' That is to some extent what Swiss did not do: it kept trying to repeat the past. But there is a little paradox: the CEO of the relaunched Swiss International was also an ex-pilot. This reminds us that the same characteristic can deliver opposite results.

To support this need to turn the page, comments were often heard regarding the new name of Swissair: Swiss. Airlines' names have been a topic for acronyms for years and Swiss did not escape the tradition, standing for So What It Is Swissair. If you do something like this, it is difficult to make the market recognize that you have broken with your own past. In SN's case, the executive customer survey quickly made it obvious to the senior executives that the market did not read SN as SABENA, because it immediately underlined the differences that the new company was embodying.

A turnaround demands a break with recent corporate history so that each individual can start to reinvent themselves first. This point is a therapist's one. Secondly, we don't have unlimited time, as mentioned earlier: we are all running against the clock and the CEO is accountable for the whole organization focusing its energy on the new goal. Consequently, anything that can represent a diversion from the ultimate goal must be avoided for the new corporation's current and future health. That is why the first 10 minutes are so crucial during a rugby match, for example. It does not mean that the game is already over, but a context has been created and it can deeply influence the final outcome. Finally, it gives a clear signal about the style that is currently being developed.

ONE GOSPEL AND ONLY ONE: 'DIFFERENT'

The need for this complete revamping is again linked to the need to create the right environment. Any CEO who is experienced in turnarounds starts from the same basic business components: competitive advantage and organizational effectiveness. But their talent consists in combining the components in a different way. The previous model, whatever the reasons, has already shown its limits. Consequently, something totally different must be invented without any confusion. 'New' or 'different' concerns the approach, the way in which things are done differently, not the challenge itself, which still consists in making the organization thrive.

In summary, the rules of the game are simple:

- The same things are done differently, more effectively.
- Different things are done, things that have never been done before.

At this stage a strong acid test is instilled, which forces the organization continually to question itself about the relevance of its proposal. Did we do it before? *or* Are we doing it differently? We knew the former results, so why keeping doing the same things? Should we expect different results? That is in fact the definition of insanity: to keep doing the same things and expect different results!

It is through this form of questioning that change finds its roots in the heart of daily operations and forces the organization to reinvent even what seems obvious. The addiction to simple acid-test questions is crucial to saving time, energy and resources. To illustrate this point, another example comes from the strategic plans that are developed in many organizations. A strategic plan regards the allocation of resources, which will modify the competitive balance in a sector to the company's advantage. We made great progress at SN Brussels Airlines with one simple test: does it modify the competitive advantage situation in SN's favour? That gave us the possibility of capitalizing on corporate enthusiasm and discovering that it is not enough if one does not permanently think in terms of impact. 'Different' is a good acid test for continually assessing whether one is acting strategically.

BEAR IN MIND

Managing a turnaround is mainly a question of execution, an exercise in which a CEO offers the right context for their staff and the management team to work at the best of their abilities. But building a context cannot be achieved by delegation, it demands that the CEO is first in the heart of the arena, and secondly that they combine the efforts of the whole company to cascade down a style of management by example. The CEO's sense of direction is one of the foundations of this context, the goal of which is to offer the organization a single point of reference for getting out of the ditch into which it has currently fallen. The underlying reality is that a radically different performance has to be achieved with fewer staff. The appropriate amount of respect must be given to those who have left the company, to sustain the morale of those who will take it into the future.

To nurture this right context, the dynamic must come from the company itself. There is a pitfall in unwittingly confusing speed with haste by wasting the energy of an unprepared team. Don't break the bounce of your organization by mobilizing it too early towards some difficult goals, but redirect its limited energy in the correct direction through 'cluster breaker' exercises. Search for 'never been done before' exercises (NBDB), the output of which consolidates the feeling inside the company that from now on it is operating in the right context. The challenge of improving organizational effectiveness then becomes unavoidable if one does want not to lie to oneself.

Managing a turnaround successfully goes beyond providing a context and funnelling energy. It also concerns the art of managing the details. SN fine-tuned a set of topics that helped support the turnaround momentum from day to day:

- Time is a scarce resource, so a sense of urgency is everywhere.
- Comfort zones are still anchored in the past. Change occurs when management understands that it is part of the problem as well as the solution. To reach this stage, the CEO must burn a lot of energy in an intense investment in communication.
- The CEO's 'silver lining' attitude is paramount and concerns all aspects of the business. This skill must spread like a virus among the senior management team, especially during the first phase when the initial positive financial results are not yet available.
- Miracles don't happen. The company cannot believe that a plan and a cash injection will be a complete cure, they merely constitute the visible part of the iceberg. The radical management transformation is the invisible part, the real and toughest challenge.
- Don't panic. The CEO needs to orchestrate the results of the effort applied by the company and their impact on the process of 'getting out of the ditch'.

- Bury the ghosts of the past once and for all. Turn the page and start writing a new one without comparing it to the past.
- 'Different' is the word of the moment. It allows managers to assess whether the recovery plan is on track or not. The more 'different' is used and implemented the better, and the faster is the recovery, because it reflects a genuine change in attitude.

PRACTICAL REFLEXES TO GAUGE SOME KEY ASPECTS AND TO DEVELOP AN AD HOC TURNAROUND STYLE

- Put your watch on the right wrist to keep continually assessing first your own closeness to your team, and secondly whether you push or you pull.
- Have you leveraged the support of your customers in implementing radical changes in order to speed up the transformation of your organization's effectiveness?
- Have you launched a good news bulletin on any available media internally? Do you monitor your silver lining style: is it spreading like a virus among your senior management team?
- Are 'different' and 'never been done before' promoted with a passion along with the challenge?
- Are you burying the ghosts of the past?

COMMENTS AND OBSERVATIONS

José Zurstrassen, founder of Skynet,
Keytrade and Keytrade Bank[26]

A
s an entrepreneur, I agree that 'creating the correct business context to give one's team members the possibility of operating to the best of their abilities' is paramount among my daily priorities. After three experiences of creating new companies, I observe that I am currently applying this principle with the same philosophy but with a slightly different sleight of hand compared to what I can read in the second part of the *Out of the Ashes*.

First, I want to make a preliminary observation about my own context. As an entrepreneur, I am dealing with 'brand new' people, providing us with fresh and outstanding talent. Our own success naturally helps attract and keep them. But our teams are characterized by their solidarity and simultaneously the fierceness

---

[26] José Zurstrassen founded Internet provider Skynet, which was bought by Belgacom, global trading company Keytrade and Keytrade Bank, which was bought by Crédit Agricole.

among their members, who are continually checking that each of them contributes as expected to sustaining the team's progress through their individual talents and skills. Our team functions as a filter, which says to the senior management or the executive committee, 'We need to find a solution for this person because they are not appropriate for the nature of the challenge they are currently involved in.' This represents in my opinion a substantial difference to the turnaround context, because these companies are perpetuating a long history whatever turmoil they go through. In my opinion, a less entrepreneurial organization is more inclined to be lenient and complacent with performance progress. Under the alibi of culture, one does not dare to address this kind of topic due to a deeply entrenched practice of not criticizing one's own colleagues. In fact, these behaviours threaten the long-term survival or competitiveness of the whole organization. But simultaneously, my role as CEO in this context is to prevent excess in order that we continue to build the corporate culture and avoid falling into uncontrolled chaos or a jungle.

Peter Davies implemented the 'context-building' principle in achieving superior results with the same people, leading them through a series of never-been-done-before confidence-building experiences. When Peter Davies said 'with the same people' I read 'with those who have stayed', which is not intrinsically an advantage. But in our small team of advanced IT experts, we continually have to overcome a risk that plagues us due to the skills and style of these people: unwittingly or not, self-limiting one's own ambition. This is paradoxical compared to what I said above, but it just reflects an inherent characteristic of human nature.

A team of IT people like ours is characterized by its creativity and its quality of analysis, but paradoxically the self-limitation of ambition results from overanalysis, which leads quite naturally to paralysis. The more one observes the target, the more it swells and the more it scares. Consequently, I have discovered that my role is about changing the paradigm for skills that have been my life for more than 20 years. To build confidence and create the right context for success, it is necessary to lower the threshold of the perceived difficulties. This must not be read as a form of management by complacency, but as management by hands-on experience in the heart of a confidence-building dynamic. If one begins to try planning everything to align one's own skills with the goal, very quickly the difficulties inherent to any project accumulate and one feels swamped by increasing complexity. So the goal will never be reached, and naturally one looks for excuses.

But paradoxically, by lowering the perception of the difficulties the project almost naturally becomes actionable. This proposal is more than a project trick, it is a management principle. When an organization is in motion, the magnitude of the difficulties is perceived in relation to its own dynamic of progress. Consequently, the difficulty can be more realistically assessed and the dynamic avoids ending up stuck and solution-less. An analogy with a sales context in a business development situation illustrates this situation. I am not obsessed by the business plan but by the first order, because as soon as it is achieved one has something tangible to deal with, for instance promises made to a customer. The right context is associated with tangibles, and tangibles are what our collaborators are naturally attracted by so that they develop a genuine perception that they are operating in the correct context. Tangible components help them develop a mental representation of their contribution or what they have to achieve. Then, it is my mission as CEO to bring this representation together in a way that will become the competitive advantage of our organization.

To support my approach, I can take the example of reengineering the IT back-office process in Keytrade Bank (www. keytradebank.com). This was achieved over three years with two people, or the equivalent of five man-years. If at the beginning we had tried to document everything, we would have remained stuck with a challenge made up of three independent back-office systems, merging into one at the end, while maintaining the service on the other two during the transition period. Outsourcing this task would have cost us 50 times this investment.

First, the result was achieved by genuine entrepreneurs who welcome challenges, as I said above. But that was not enough. The difference lay in the entrepreneur's pioneer mindset. That explains how we could cope more naturally with uncertainty in this challenge, but in my opinion the picture was even then not complete. Among pioneers there is a specific breed, the pathfinder, who doesn't need marks on the ground to organize their progress. It is this characteristic that explains the differences and this that we nurture through a specific practice to benefit from a breakthrough. In our corporate culture we are relentless at eradicating anything that does not lead to the optimal solution. Everybody agrees with that, but we have become expert in what spoils the optimal solution.

First, we do not let analysts 'fall in love' with their analysis, we are solution oriented. So we don't want paper, we want to hear how they have already solved the problem in their head. The test is very straightforward: when the description on the blackboard is clear, fast, to the point and sharp, we have the signal that they are ready to write the software. This is what is called extreme programming. So in our culture we don't write software straight away, we come to the blackboard in front of peers who play a devil's advocate role. We are not here to validate, we give confidence to our team members that they are empowered and accountable, they can move ahead. Frankly, sometimes we do not understand the detail of what they have in their mind, but it clicks. Any area where an analyst is not at ease is a potential area where a few months down the road there may be a bug, meaning that they have not fully understood what to do and consequently we waste money. In this case they have not made the effort to make it clear enough either for us or the customer. We promote this systematic effort because it is a means of avoiding sharing problems in an ineffective way by simply adding resources, instead of diving into the depth of the issue to formulate where the bottleneck is. Our blackboard is our ad hoc entrepreneur complacency repellent. The effective turnaround manager is driven by the same objective of repelling complacency – the difference, in my opinion, is just the chosen theme.

To manage a team of pathfinders, I need to keep them continually on the alert and exploring faster than anybody else the relevance of a possible track, with the main goal permanently in front of their eyes. We constantly and simultaneously move and keep analyzing. This tempo is my responsibility. It is a form of management by example, where I need to operate among my team members with my shirtsleeves rolled up. To obtain superior results from these people, closeness is critical. But that is not enough in the heart of the challenge. Closeness combined with a feeling of expertise- or skills-based respect must also emerge. It is this combination of closeness and expertise that naturally leads to the sense of a tailor-made, stretched goal when one observes the process from the sidelines. In the heart of the dynamic, everything is far more natural, more spontaneous, more actionable.

At this point, I want to return to the observations made in this chapter regarding not wasting resources on inappropriate objectives. I encounter the same challenge. Our style is to implement a never-ending commitment until the end of the project. But so that the team can keep developing this precious mindset, both the recovery period and the art of giving oneself a break are crucial. It is why my team members must also be great at doing whatever they want, even nothing. By nothing, I mean without an objective of planned return.

'No objective of planned return' is a true form of organizational discipline to keep moving ahead in a context where the solution seems to be out of reach. At some stage, so much information has been accumulated that is very difficult to find meaning in order to move ahead. From my experience, this is the signal that one must give oneself a break. My closeness to the team is a great advantage in suggesting the break, even sometimes imposing it. But this is also a means of reinforcing the team's confidence in themselves and in their management. It is interpreted as such by my team even though I am permanently monitoring their rhythm.

In our company, a break takes two forms:

• A structured one, which is a milestone of our corporate culture. Each year, two dialogue sessions are organized just so we can speak together outside of the office on any topic except the current business. It is our cluster-breaker approach. This is crucial, because the team of entrepreneurs at the early stages of the adventure have to welcome the professional managers who are preparing the company's future. And they need to know one another.

• The second form of break is less structured, more spontaneous. It represents a form of individual discipline through which one gauges that one is stuck and cannot go further with this dynamic. So a rejuvenation break becomes necessary. But it is important to note that this need for a break is not a form of self-limiting attitude nor a form of waste, it is a form of advanced entrepreneurial management. This attitude makes us move to another layer of management philosophy. The approach is a means to plough and replough the field. In order to draw benefits from all these management attitudes, the company culture must promote an open mind.

Finally, with this break human alchemy begins to operate. But in this case one must be ready to react appropriately. Given that we were stuck, when the first shoots of the solution occur we must not be surprised if they are not growing where we expected, because we had no clue. We have ploughed a field with our best

professionalism and talent, but we must also welcome opportunities, whatever they are. We cannot say 'It is not what we were
expecting', we have to ask 'What does it mean?' It is through this
set of management principles

- Lowering the threshold of difficulty by moving ahead without
  any delay.
- Being ready to give oneself a break to allow a natural rejuvenation process.
- Continuing to wonder about the meaning of the weak signals
  of any opportunity.

that I nurture the right context to give a future to my organization. As is addressed in more detail in the case of SN, these are
people-centred management practices, a topic where one keeps
learning every day.

PART III

IN THE HEART OF
THE TURNAROUND

*I*n a turnaround the diagnosis stage is of paramount importance, but if it is achieved with tools imbued with the company's
past, then there are two risks:

- First, the team will not comply for the simple reason that they
  think they already know what to do.
- Secondly, the team won't be motivated and will think that
  they can't do any better than before.

From experience, even in the diagnosis phase I consider it crucial
to bring in a new way of dealing with the turmoil of the company's daily life. A well-formulated diagnosis is already 80 per cent
of the solution. But reaching this stage supposes that the company
can immediately observe a difference both in terms of the approach
to its problem and the tools being used. Therefore, in the early
stage of the turnaround it is wise to distrust the tools and informa-

tion systems that are already in place. They have contributed to leading the company into its crisis, so it is a fallacy to imagine that using them without any in-depth adjustments won't replicate the past.

In a crisis context, the company is surrounded by an inextricable confusion between the causes and the symptoms of the problem that is plaguing it. Moreover, it is almost impossible to prioritize the causes, which leaves the company more and more unable to react. It doesn't know from which angle to attack its recovery effectively. Therefore it enters into a form of collective incantation, where the whole corporate body becomes infected and repeats 'When we receive the $300 million equity injection from the government, the problem will be solved', as in the case of Caribbean airline BWIA in early 2006. This is hardly correct, because the company and its staff have lost track with the complexity of their own problem. For instance, in this technically bankrupt company the causes of its endemic problem were almost inextricable for the staff, but the reality was actually quite simple. In 2004 the issue was overheads related, then in 2005 it combined with an explosion in direct operating costs, and finally in 2006 the problem increased in complexity with a decline in the number of passengers. Consequently the company was nearly paralysed.

When one jumps into the heart of a turnaround, the nature of the exercise mainly consists in changing the angle of the light by implementing a series of different management tools that form part of a new toolkit. This offers a silver lining in a rather blurred context. In this third part, through experiences drawn from projects in different industry sectors as well as in SN, I propose a three-step approach to jumping into the heart of a turnaround:

- First, free the company from its performance measurement 'derivatives' that are rooted in the past.

- Secondly, be absolutely determined in reimplementing basic management tools to find a rejuvenated way to address performance measurement.
- Finally, revisit the fundamentals of the business model at the elementary level of strategy execution and be disruptive in managing performance measurement in the company.

CHAPTER 6

APPLYING THE WORD 'DIFFERENT' IN THE AREA OF PERFORMANCE MEASUREMENT

THE TRAP OF PERFORMANCE DERIVATIVES

In both consulting and research assignments, I have observed that companies are keen on delivering thick performance reports. Discussing with managers how they use them, one notes short cuts, justified by efficiency, which they guarantee as their infallible way to assess how performance is going almost on the spot. These short cuts have a name: performance measurement derivatives.

A performance measurement derivative is a means of measuring performance that is technically correct, but that encompasses a high level of abstraction. Its oversimplification runs the risk of disconnecting everybody in the organization from a clear sense of their contribution to the final result and their possibility of reacting appropriately. Consequently, it is a correct representation of one aspect of current performance, which describes the performance but not how it has been reached nor what the next step is.

For instance, in the airline sector two well-known performance indicators are 'rask' (revenue per available seat kilometre) and 'cask' (cost per available seat kilometre; see Figure 10.1), which one encounters as soon one crosses the threshold of an airline company. But each sector develops its own indicators, as Business story 6.1 demonstrates.

Business story 6.1: 'I just look at the white paint'

Fond Partenaires, the capital development division of Lazard Brothers, in the mid–1990s bought a decorating products chain, which they developed by acquisition to transform it into a leading distributor with more than 300 shops in France and Belgium (there is more on this story in *Organic Growth*). The core of the business initially was wallpaper. The performance management system of the wallpaper outlets was very well designed, just one sheet of paper with the critical information both for the shop manager and the supervisors.

When customizing a senior management seminar for the company, I was having a discussion with a regional director in his car parked just in front of the main window of the shop. I was listening to his comments on how the retail marketing efforts were cascaded into the network, when he saw on my lap the shop's performance sheet. 'There is a lot of data and it is difficult to make a synthetic analysis before you enter the shop, so you need to rely on one indicator to assess nearly on the spot how the shop is performing,' he said. 'The volume of white paint sold is my key. The reason is simple,' he continued, 'when somebody changes the wallpaper, they are also repainting the ceiling.' So the sale of one accessory indicates the performance of the shop. This may be effective, but is it reliable enough to manage and mobilize the energy of the shop manager? One can easily imagine a lot of bias in such an approach.

When I checked with a senior manager the appropriateness of this technique, he smiled and said, 'It used to be one of our tricks at the beginning, but now I wouldn't recommend it. I don't know if the correlation between white paint and wallpaper volume has remained the same due to the evolution of the pattern of consumption and customer behaviour.'

So derivatives are acceptable, but over time they might become misleading. Today's success prepares for tomorrow's failures, a point that also applies to the way performance is measured. Many other sectors have also underlined the risk of this form of intellectual approach. This makes you run the risk of missing opportunities by staying stuck to an inappropriate paradigm. Coca-Cola's famous 'share of stomach' turned into a monomaniacal focus. Due to this kind of culture and practice the development of bottled water was considered as a big-volume business distraction, which it was not. Ask Evian its opinion! Organizations caught in this trap first don't consider with enough attention the weak signals of the market, and secondly when the executive side is conscious of the need to organize a reaction, then the board often vetoes any strategic shift.

In any organization, performance derivatives are associated with a particular era. When a new collective mindset has to be developed, the page must be turned at the same time. The previous business model's performance indicators must also be left behind, or at least their basic relevance revalidated. Organizationally speaking, performance derivatives are brilliant but only for those who invented them, who have usually left the company long ago. Moreover, they are too often loaded with a great deal of emotion from the previous era. Consequently, it is wise to think differently and even abandon them. But then the question is what to substitute for them. Before thinking in terms of content, I suggest first reflecting on the characteristics of performance measurement, which helps avoid the above traps.

A revalidated performance measurement system has two characteristics:

- It is an 'evergreen' concept as opposed to a multilayered evolution. Too often ad hoc analysis becomes the backbone of an information system, but being ad hoc in analysis is paradoxically not very flexible. The analysis is technically correct, but links too much to the specific nature of the business situation. An evergreen concept is not evergreen because of its results, which are just final fine-tuning, but because of its fundamentals. This means that this system of performance measurement is always refocusing attention as a starting point on the ad hoc essence of a specific business. Then one can be creative in analysis, but with a substantial difference: one starts off on the correct foot.

- It is accessible to the majority of the company. In relation to this characteristic, it is not just a question of how sophisticated the users are, but of up-front investment in the simplicity of the tool's design to observe performance while keeping both feet on the ground. Consequently, what has to be observed? This is defined by the key steps of the business model. I have heard a lot of people speaking about business models, but too often and too many of them missed its key output in terms of learning. A business model is not the sophisticated result of an (expensive) consulting study, but should be the outcome of a consistent development of thinking. That is why my first task is always to try to figure out a business model on the back of an envelope with simple tools such as pocket money analysis, a waterfall or break-even point. All these graphs have one key characteristic in common: a sense of hierarchy, sorting out what is essential, what is less necessary and what is driving the business. So the graph tells you immediately where to focus your attention.

When performance is assessed through derivatives, very quickly the business model or the fundamentals of performance are forgotten. So one starts limiting one's own observations to the results, but not to the relationship between the causes and the effects. Analysis is about finding links, or in Latin *interlegere*, which incidentally is the root of 'intelligent'. Through this analysis one is looking for an intelligent system or fluid system of links that is accessible to the vast majority, because it is this vast majority who need to be mobilized. Consequently, the system develops its own blinkers and little by little the real meaning of the results becomes blurred. Paradoxically, the company ends up lost in correct data.

Even worse, the company could also develop reflexes that are misleading or at least lead it to miss opportunities. This is followed by a form of corporate inhibition over proactive initiatives. The reason is simple: the derivatives at a sector level are used by the whole sector and this leads to stereotypical behaviour, which reinforces the position of the leader and prevents the development of challengers. True challengers' performance measurement indicators confirm both their status and their competitive value.

I have had the opportunity to observe the characteristics of challengers quite often. The story of Holland Casino in Business story 6.2 illustrates how real challengers not only challenge the performance indicators but also the assessment of their real impact.

> **Don't give a hand to the sector leader's competitiveness.**

Business story 6.2: 'We welcomed twice as many players'

In 2002, the city of Brussels was allowed to open a casino downtown due to a change in the federal law in Belgium. Many casino groups competed to get the licence for this new project. One of the competitors, Holland Casino, was a real challenger. It was the first time this company was applying to invest in a venture abroad. Up to then, it had just offered management contracts. This time, it would be an equity operation if its proposal succeeded.

Why should Holland Casino be considered as a challenger? There were many reasons, but one particularly struck me. The whole sector formally communicates the average spend per player. In contrast, this challenger was the only one to devote a full section of its annual report to the number of players who crossed the threshold of its casinos. Is this important? Critical! It means that its real concern was the combination of three questions:

1. How I am attractive?
2. Do I keep recruiting new customers?
3. Is my entertainment concept effective in achieving this?

This company behaves like a distributor, which in fact it is: a distributor of entertainment. Finally, just to give a flavour of how the challenger is changing the rules of the game, I would like to report a conversation that took place during the negotiation of the project. The topic was to try to benchmark this organization's performance against another highly visible western European operator, Casino Austria, which runs the same number of casinos in its home country, welcoming 3 million players and achieving the highest average European performance in terms of spending per player. The question was about this 15 per cent difference in terms of spending per player and how the challenger could explain the superior performance of the leader.

Its answer was devastating. 'We must confess,' said the head of international operations, 'that they are far better than us in this area, we made progress in this field, we are 15 per cent behind them after 3 per cent progress last year. We carefully monitor the evolution of this ratio and many of our plans aim at filling this gap.'

'But maybe another piece of information would interest Professor Mognetti for his partnership assessment,' said a representative from Holland Casino. 'We welcomed twice as many players to the same number of casinos . . .'

Who would you choose as your future partner? There is a real risk of following the sector's gospel like lemmings. Consequently, always keep an eye out for disruptive ways to look at what seems obvious. Both natural arrogance and confidence in one's own routine, which is a great form of complacency, make assessing performance a very demanding task, which does not guarantee that one enjoys its full impact. In many cases, one just misses the potential impact and disruptive effect of that assessment. Consequently, this topic of renewing performance measurement indicators is a strong reminder for the CEO − and for the whole senior executive body − of their sense of weak signals and the time they spend outside their office dealing with creating the future. Performance measurement is a general management topic, it is not a question of expertise. It brings the management committee into the heart of the competitive arena, because derivatives reinforce the current leader's position and stifle the potential for creativity. To break this vicious circle the initiative must come from the top.

> **A true challenger's performance indicators reflects their real status.**

In SN the objective was quite clear: 'rask cask' performance derivatives were technically correct but not appropriate for the time being. It was not with these indicators that one would focus the corporate energy on a new confidence-building process that fitted SN's strategic mission to 'Connect Brussels'. Something more relevant needed to be developed.

AVOIDING THE TRAP OF PERFORMANCE DERIVATIVES

The warning of the risk associated with performance derivatives is clear, but the question now is what kind of management lucidity avoids an organization falling into this trap.

Not relying on performance measurement derivatives requires continual questioning of what is behind the performance derivative information. It is a very demanding management development approach, the real challenge of which consists in shifting from understanding to assimilating. But how does it become a corporate reflex? Through the following short example, I assessed that in SN Brussels Airlines this transformation had effectively occurred.

> **Diagnosis puts it into perspective – easier said than done, but paramount.**

An airline is a costly deployment of assets. Consequently, the profitability of a route is influenced by a correct fine-tuning of how closely, on a specific route, the number of flights or the deployment of assets matches the legitimate accessible market, the potential revenue. On a specific leisure[27] route to Italy, SN Brussels Airlines encountered some real difficulties in achieving the desired profitability due to many factors. In low season (winter, from November to the end of March) the performance track record was an operating loss, so for the next year's low season some measures were discussed and immediately implemented. Unfortunately, the same operating loss was reported one year later. If one limits the observation to the return on direct operating cost – RoDoC, in corporate jargon – this indicator gave the same poor figure. But it is paradoxical in this context that there was a positive signal of an appropriate management evolution in the correct direction.

Erwin, one of the planning managers, dashed into his boss's office. 'Christine, we have a real problem,' he said. 'The RoDoC is the same with half as many flights. So the reality of the performance is even worse at the corporate level, because this figure did not reflect the non-absorbed fixed cost of the grounded aircraft. We must return to the former frequency of flights, because I think that we have certainly reached a level of service where our offer may become inappropriate for the market. So I suggest observing the January results and taking a decision to implement some corrective measures in early March.'

By his comments and attitude, Erwin was clearly showing that he was mastering the business model at the elementary level of strategy execution for which he was accountable and behaving with the expected level of autonomy. That is the guarantee of

[27] A leisure route is characterized by a dominant traffic of tourists or people visiting friends or relatives. The potential of business travellers is limited. Consequently, the average yield or fare paid by the passenger is not as high as on a business route where travellers are ready to pay more for more flexibility in their flight options.

keeping the company on the correct performance improvement track.

This example was for me one of 1000 weak signals that every day this company was developing tangible evidence that something different was brewing. This starts by smart executives' reactions, those who are able to reposition their diagnosis with respect to the main picture and suggest that a quick decision be taken. The next question is: why can you observe this? The context is always very important. SN had been relaunched with 1800 traumatized people. To obtain this kind of result a lot of hurdles were already behind them. This executive had become more confident in the style with which he played the game. The inhibitions had been cleared.

By introducing a new performance assessment model, individuals are strongly prompted to wonder about the real meaning behind each result. But it is not enough to announce a new and more effective approach, potentially leading to a form of stealthy competitive advantage. The means of supporting this permanent requestioning must also be addressed from the beginning. In the above situation, the senior executive's style encompassed a combination of 'open door' and 'empathy', which created the right context. Christine can genuinely listen, with a real commitment indicating that she cares. But then comes the next critical question: who will take the initiative of checking what is missing from this diagnosis?

Subordinates' new skills guarantee good suggestions – senior management's style prepares the field for the next ten suggestions.

Good ideas without any serious follow-up never produce any substantial progress. The executive who prompted the discussion will not miss the deadline for producing results: it is their own priority. But it is now the responsibility of senior executives to send the message to their subordinates that they care about them as much. The reason for this combination is simple: to keep ploughing the field to create the correct context for the next ten suggestions. It is the responsibility of the senior manager to leave their office to wander around and secure the next step of the diagnosis-correction process. New management reflexes and analysis are mandatory, but style is crucial to prolonging the momentum.

CHAPTER 7

A NEW MANAGEMENT TOOLBOX

BASIC TOOLS

Freeing the company from performance derivatives rooted in the past avoids it falling into the trap of not having the discipline to focus systematically on what is behind the result. We have observed a new dynamic of performance measurement in action, but the question is what has led to such a result. The next step will consist in illustrating, based on different business situations, which management tool will support this transformation.

The performance measurement tools developed or used in the SN case are not distinguished by their sophistication. Their key characteristic is to be deeply rooted in the common sense of this specific business. This sounds obvious, but paradoxically it is where the challenge begins when one tries to reinstil the obvious elements. This task is not as simple as it appears and many difficulties are encountered. People think they know these obvious things already, so there is a real risk of boring a senior management audience or a whole company when talking about them. But withdrawing from these pressures runs the risk of building a new system on foundations that have not been sufficiently validated.

Sticking to the basics and being aware of the fads is a difficult exercise. Very often it is taken for granted, so the challenge is somewhere else. This is partly true, but with which toolbox should one address this challenge? This is not often the appropriate question. When in December 1996, nearly 15 years after the publication of his book *Competitive Strategy*, Michael Porter wrote an article in *Harvard Business Review* entitled 'What is strategy?', this article was a crystal-clear and necessary clarification of the risks run by many organizations in not recognizing the bias that they are developing in their strategy. In management the issue is rarely a question of being wrong, because the correction then is obvious, but of deviation, which requires more time just to recognize it.

In a turnaround or a relaunch, the challenge is not about being brilliant but about developing the smallest possible number of handicaps. For this kind of contest you don't need a racehorse but a mule. This is a very solid animal that perfectly fits the challenge of riding along a mountain path. So to reach the objective you need to make the smallest number of mistakes, not have the most impressive pedigree. A mule will never make a mistake such as putting one of its feet into a hole, while this will happen with a horse. But on the other hand, to make permanent progress towards the objective demands a specific sleight of hand, if one does not want to be caught in the famous image of a stubborn mule that does not want to move any more.

> **Basic management techniques are scary, but that's not a reason to take them for granted.**

Consequently, the next question is not whether a management tool is good or bad, but whether it is appropriate to the context. Given that in a turnaround the organizational skills are under redevelopment, the warning is clear: don't put the cart before the horse (or the mule!). Redeveloping the skills of an organization is the same kind of challenge as developing those of young future talent through an MBA programme.

Business story 7.1: 'What have you done with the basic test?'

At the 20th anniversary of an MBA class at IMD in Lausanne, some alumni were involved in an open discussion with the current MBA programme director regarding the evolution of the content of the curriculum. The MBA programme used to be based on collective and individual exams. After a couple of years of experience, one of these exams was systematically recognized by the vast majority of the participants as the most relevant during the entire programme. It was called the 'basic test' and MBA students had to pass this test in each teaching area. It represented the rock-bottom level of management techniques, tools and knowledge that all MBA students must master to expect one day not only to graduate, but to address a business situation with both insight and common sense. The philosophy was the following: before giving any form of dominant flavour to one's own study, it is important not to develop any kind of handicap from the beginning – in other words, you have to start on the correct foot.

Some alumni asked the current MBA programme director how the basic test had evolved. The answer was: 'We got rid of it.' Then he started to outline the new leadership orientation he was giving to the programme. That was certainly appropriate, but it was interesting to observe the exchange of questions in the eyes of the audience, wondering whether he was not underestimating a crucial step in the development of a manager. I personally thought that this was the case, which proves that this kind of problem can occur even in the most sophisticated management development organization.

The lesson of this case applied to a specific context of turnaround like SN Brussels Airlines is: don't take basic management tools for granted. If you do, you risk losing contact with common sense and daily business reality. Managers lose ground and the sustainable effects of the turnaround move further away.

BASIC TOOLS ARE IRRITATING!

Addressing the question of basic performance measurement tools, you hear the same from everybody: 'We already know that.' At least it is a good, positive start to hear this. But something being known does not mean that it is understood, or assimilated.

In this challenge with basic tools, I even encountered some difficulties with my own staff. In September 2002, for four months a team of three of my students on the HEC MBA programme joined SN for a company consulting project.[28] They perceived this turnaround context as a unique opportunity to experience part of the content of the MBA programme. Quickly on the campus in the south of Paris a rumour spread: 'This Mognetti is incredible,

---

[28] I coached more than 100 MBA company consulting projects with clients in Europe, US and Asia, both at HEC and at ESCP-EAP from 1987 to 2002. In my opinion, this is one of the distinctive characteristics of a MBA programme, being plugged into business reality. I must not forget to say in a form of testimonial that the correct and enthusiastic tempo was invented by Professor Ian Kubes from IMD. I adapted and leveraged what he taught me. I adapted it to my specific context, but simultaneously stayed as close as possible to his unique pedagogical insight in this field.

he is focusing all our efforts on break-even point analysis. It is ridiculous! We are here to develop a growth strategy, not waste our time with obvious things.' First, nothing in management is obvious; and secondly, they substantially contributed to preparing the context for a growth strategy. They did not realize that it would take more than six months to transform the concept of break-even point into a corporate reflex. Reactivating good old things may be very effective, but the investment must not be underestimated. In fact, one starts all over again almost from scratch with concepts that are only individually familiar. But a brand new management approach has to be written for the first time. Among the MBA participants involved in this project only one, a brilliant Peruvian woman, came back one day from a meeting with a group of SN staff with this salient comment about the meaning of an MBA management challenge: 'The danger with an MBA programme is that we too easily consider that it continues into our first job. The study group is over and the art is to obtain outstanding results with the available resources. It is a real management challenge.' Consequently, 'basic tools' are a very difficult topic because of the apparent ease of approaching or understanding them.

> Don't confuse speed and haste –
> management fads worsen the trap.

It is fair to bear in mind that creating the right context always encompasses a technical and an educational dimension. It is a serious concern, and cannot be satisfied with intermediate performance on these basic techniques. One cannot tolerate approximation in this form of management reflex. If one does, due to one's own complacency, the whole organization runs the risk of starting off on the wrong foot. And this occurs in situations of the highest tension. A fatal error will be made in ruining the organization's speed of reaction. Finally, one is lying to oneself, because corporate confidence follows individual confidence, so how is it possible to achieve it if this process starts by making concessions on fundamentals? Thus the first step is to reconfirm individual confidence in these basic performance assessment management techniques to avoid developing handicaps that would inevitably penalize the next steps of the recovery process.

> **Don't accept trade-offs. The goal is to democratize the monitoring of strategic execution.**

Confidence in one's own skills is critical, but the purpose is not to pass a test, but to create together the future of the organization. This again leads to a question of choosing the appropriate tool out of the corporate toolbox. For instance, creating a business unit profit-and-loss account is an important step in progressing the evolution of an organization's performance assessment. It is often presented as clear evidence of an organization's progress, but it is generally more related to accounting.

THE COMMUNICATION CHALLENGE

With the basic management tools, which I always have in mind, the bar is raised for the whole corporation, up to the heart of something more challenging: creating the right environment to democratize the monitoring of how the strategy is executed. This has to be achieved with simple tools because it must set up a dialogue with the vast majority of the organization. But that does not mean it will be easy to reach, any more than the word *simple* also has to encompass *appropriate*.

Business story 7.2: Counting in physical units

My idea of democratizing the monitoring of strategy execution is deeply rooted in my academic experience, during which in the mid-1980s I came across Ford of Europe. In the early 1980s it was assessed as one of the best-managed automotive companies. Its management control techniques were considered the source of its competitiveness. But a note written by one of the company's auditors at the time reported that its efficiency was being outpaced by a fast-emerging Japanese automotive manufacturer, Toyota. The latter had developed a trick in its management control approach, which was allowing it to achieve the same level of control with the same accuracy and effectiveness but nearly three times fewer controllers. The answer to this conundrum was simple: the 'corporate currency' used was not the same as at Ford. The Japanese 'currency' was deeply plugged into the daily reality of the business, while Ford's was too abstract − it controlled with Wall Street-based currency, the dollar.

Consequently, in my different consulting assignments I have promoted performance measurement systems that mainly rely on 'physicals' instead of on euros or dollars. You can always turn a measure into a value, but it is crucial for controlling the execution of a strategy to broaden the base of committed people who know that they are accountable for what they see, touch and count, and not for what is subject to interpretation. Interpretation of the data is the responsibility of senior management. Finally, using physicals offers a great opportunity to share the performance to a deeper level never reached before for a very modest cost.

This culture of physical-based performance measurement produces another benefit: it forces the company to stay focused on volume. The next story helps support this proposal.

In 2002 I was asked by the CEO of Kenzo, a garment brand belonging to LVMH, to help him prepare his strategic plan for 2003–2005. Commenting on the ongoing plan, the CEO explained that between 1998 and 2001 sales had increased by a particular amount, which was a good rate although maybe not the best in the division. Then he spoke about collections, trends, competition, but nothing about the physicals, such as number of customers per shop, number of garments sold. I discovered the next day in discussions with a young controller another business reality of this company: over the same period where the company had grown in sales volume expressed in euros, the number of garments or units sold had decreased by 20 per cent.

This was strategic and very annoying: this company was selling fewer more expensive products to fewer customers. So the question was simple: how long could such a trend last? This company was imperceptibly losing touch with business reality. In fact, I arrived too late because the parent company had made the same analysis as me and the future of the CEO was already settled outside the company. My contribution just deferred the decision by three months.

By not looking at the business through physical measures, senior managers narrow their in-depth understanding of how the business works. Business story 7.3 reinforces this point.

Business story 7.3: Taking the physical pulse

A second example is even more illustrative of this interest in physical measures as the means to 'take the pulse' of a company. Above I discussed Fonds Partenaires, the owner of a house decoration products chain. The wallpaper shops were called Chantemur.

I explained in *Organic Growth* that around 70 per cent of the market in France is controlled by two chains, Chantemur and Quatre Murs. Both chains initially were family businesses, and the founders were cousins. In the mid-1990s the owner of the Chantemur chain proposed a management buyout to its senior executives, who successfully took up the challenge. This buyout was a spectacular success: the debt was paid off in half the time that had been planned.

Then a second buyout was launched with a different financial institution, Fonds Partenaires. The debt being paid, now the goal was to grow through organic growth and acquisitions. This second buyout went well too. But when I started to advise the CEO of this company, it was already sending some signals that it had lost a part of its focus on its traditional business fundamentals. Thanks to the family relationship, the two companies kept exchanging their respective performance in detail, shop by shop. It was a kind of family PIMS.[29] Another characteristic of this business was that both brands were present in the top 100 main cities of France. Consequently, the shops were competing against each other. For each shop, one sheet of paper delivered to the management of both companies in two columns and a few rows the product segmentation by price, the number of units sold, sales turnover per category and total sales turnover, plus some other data regarding the accessories business (paint etc.). It would be difficult to have a better vision of the market at the elementary level of strategy execution – the large French cities.

When I discussed with the executive committee running the buyout how they assessed their relative performance compared to their 'cousins', the comments were very positive. They

---

[29] PIMS stands for Profit Impact Marketing System, a 1970s database where subscribers pooled their data in a hidden way in order to achieve marketing effectiveness benchmarking.

suggested that I should dig further into their business with the help of the details of these shops' relative performance. The reports were missing some form of consolidation, but that was easily solved. This naturally produced some conclusions, but they were unfortunately somewhat different from the enthusiasm and confidence of the executive committee members. The business was good, but not in relative terms. In sales turnover per shop, the performance remained on a par with Chantemur's although there were some annoying signals. The sales turnover was the same, but little by little Chantemur was beating its cousin in the average sale price of rolls of wallpaper sold, so Fonds Partenaires was losing ground in volume. One of the cousins kept focusing on the number of units sold, while the other was focusing on the average revenue from the units sold. This is bad news when you are a category killer. Stuck in buyout mode, the former distribution executives had sold their volume shopkeeper soul to a buyout operator one, so part of their basics had been lost. It is a volume business and beating the 'coopetitor' on average price is an analyst's perspective. In summary, they had the icing, but without the cake.

Was it a problem at that time? When I observed this situation eight years ago, it was not. The executive committee and the CEO listened to me with some interest, but my report did not create a revolution among the members. Only one commented on my analysis, saying, 'It is very annoying because we have here a signal that we are less and less able to sell to the "proletariat".'[30] At that time, this person was not among the higher echelons in this committee, so his comments were considered as rather a joke and not worth any additional consideration. Unfortunately, he was 300 per cent right!

---

[30] This was a very strong but correct image to express the heart of their market positioning of the company. Their decoration solutions were very cheap, DIY ones.

There is an epilogue to the Chantemur story: finally Lazard sold the group in sections, and the survivors of the initial team made another buyout on the wallpaper side. While I was preparing this book, I had a discussion over the phone with the president of the executive committee, who told me that it was very difficult because they had to reinvent, rediscover and rewire the business fundamentals, which had remained unconsidered for a very long time. So not looking at the business performance through physicals meant taking a risk. God is in the details, said Goethe, and that applies to good management too – the only question is how we talk about the details. Due to its success, the team developed a form of management complacency not only in its discipline of performance observation, but also in its capacity to react, implement and monitor corrective actions. But on the other hand, the performance conformed financially to what the private equity firm was expecting. Consequently, they mistook the trees for the forest and started missing the essentials, the physicals that expressed the quintessence of the business.

When the genuinely physical is abandoned for something more sophisticated, the risk of management mistakes increases for two reasons. The senior manager or the CEO:

- Is little by little disconnected from the business reality and from the vast majority of those who struggle to generate in the details the quality of daily performance. Moreover, this situation is perverse because it happens imperceptibly and unwittingly, with a mistaken feeling that they can catch up almost on the spot.
- Is cocooned in the most dangerous place from which to see the world, their office.

In contrast, these who have these physical measures constantly updated in the back of their mind are plugged into the heart of their respective businesses. One can observe that these people, ranging from a shopkeeper in a distribution chain to a CEO, possess one characteristic in common: they know how to anticipate.

In the management toolbox development process at SN, my suggestion was twofold:

- First, adopt the principle that you need to walk before you can run. Consequently, do not sacrifice basic management tools or practices for more sophisticated versions, because the turnaround manager needs these to be easily understood in order to motivate staff. It will be their responsibility to control how the tools are assimilated.
- Secondly, keep developing this ease of access to performance information by the use of the correct physical measures.

This management tool review process is a preliminary step heading towards revisiting the company business model, in order to share with the whole team the overall picture and its drivers, representing the future pitch on which to execute the turnaround.

CHAPTER 8

THE BUSINESS MODEL REVISITED

THE ELEMENTARY LEVEL OF STRATEGY EXECUTION

A business model is the expression of how a company has chosen to make a profit in its sector. This tool is for business 'doers', and consequently its management value must be observed at the elementary level of strategy execution. It is also at this level that the basic tools approach evoked in the previous chapter will help the doer make the business model thrive. The turnaround process enters into its company transformation phase when, in the specific corporate context, the elementary level or levels of strategic execution are validated so that the relevance of the business model can be shared. This is what was researched in the first few months of the SN project. The purpose of this chapter is to present the different steps of this new tool development process.

A business model monitors the principles of achieving performance at the elementary level of strategy execution. The elementary level of strategy execution is a result of segmenting corporate operations with respect to the strategic ambition. At this level, both allocated resources and results can be assessed with the highest level of homogeneity.

Developing a new business model is always a more ambitious task than it appears. The business model reflects the selected strategic mission. In SN Brussels Airlines' context, the strategic mission was simple and was given to the Chairman and the CEO by the board: 'Connect Brussels'. That precisely underlines which elements had to be scrupulously monitored from a strategic point of view.

The mission statement clearly says what business was *not* for SN Brussels Airlines. For instance, SN did not have the ambition of promoting a hub in Brussels, which was a very specific strategic positioning. Simultaneously, this meant that it would exclusively focus on point-to-point business. Some connecting business would also be legitimate, but this traffic would be a complementary activity.

What was strategically crucial could not have been more specific: the connection, which in airline terms is a route. Then there was a consolidated level of observation: the portfolio of routes, a network. A portfolio of routes raises the question of its homogeneity or its segmentation between short-haul and long-haul operations. Consequently one is dealing with a system with four levels, like a tower of Lego bricks:

1. A flight.
2. A portfolio of flights with the same family of flight codes, a route.
3. A portfolio of routes of the same nature (short–medium haul, for instance), a network; the network is just a consolidation of a portfolio of routes.
4. A portfolio of networks, a corporation.

What had to consume the vast majority of the energy was the elementary level of strategy execution – in this case, the route with its portfolio of flights. The other levels were less strategic and represented a consolidation of the elementary level of strategy execution.

In one of my previous 'network' missions at DHL, the elementary level of strategy execution was the same. The equivalent of the route was the station, because it was at this level that the consideration of the customer was occurring and was consequently managed. The house decoration product chain Chantemur–Heytens faced exactly the same challenge. The level of monitoring of strategy execution was the shop and its dedicated catchment area. The driving philosophy of this approach is that good performance in A does not guarantee good performance in B, and poor performance in C does not spoil good performance in D. So the result is a plus and minus consolidation, and the consolidated result must never be assessed with complacency. A network is like a bicycle with a fixed gearwheel – you cannot stop pedalling, there is no rest. There is always one of the elementary units that needs intensive care.

This research into the correct level of strategy execution is not complex, but it demands some thinking to avoid errors of inappropriate observation of performance. For instance, in SN the short-haul route was quite obvious: a connection between Brussels and Geneva. In this context, the point of observation was quite natural. The performance on the Geneva route did not influence the performance on the route to Birmingham and vice versa. But in Africa the story was radically different.

Take an example in West Africa. One of SN's Airbus 330s left Brussels at 10 a.m. to land in Dakar in the afternoon, then took off 90 minutes later for Banjul in Gambia, to be back in Dakar around 10 p.m., to take off again around midnight to return to Brussels the next morning at 6.30 a.m., and after some maintenance operations the same aircraft or a sister took off again at 10 a.m. for Dakar via a tag flight in Conakry in Guinea. Africa looked like a maze of routes and identifying the strategic level of elementary observation was not so obvious. In this case, I suggested identifying what was stable as a relevant level of monitoring the strategy execution. In this example, the entry point in Africa, Dakar, was the answer. Around what we called this 'gateway' we could concentrate all the direct operating costs associated with the operation in the area. Therefore the gateway was delimiting a homogenous perimeter of operations, which could become comparable with another gateway. It would be at this level that we would concentrate all the analytical efforts. SN's African network was made up of three gateways after the crisis in Ivory Coast: Dakar, Kinshasa and Nairobi.

In summary, identifying the correct level for monitoring the execution of strategy requires an ad hoc answer to be validated in each case. The reason for the care that I suggest needs to be taken to identify this correct elementary level is driven by reasons of organizational transformation. Beyond any elementary level of strategy execution, there is always an accountable person for whom all these tools are developed in order that they achieve superb performance to the best of their abilities, first for their own pride and confidence, secondly for the whole corporate community's pride.

DRAWING OUT A BUSINESS MODEL

It is at the elementary level of monitoring strategy execution that the relevance of the company business model operates. Moreover, it gives a practical meaning to the term 'business model', which helps the business doers to concentrate their energy on what is critical in performance generation.

Some 25 years ago when I was doing my MBA at IMD in Lausanne, one of the prominent faculty members, Xavier Gilbert,

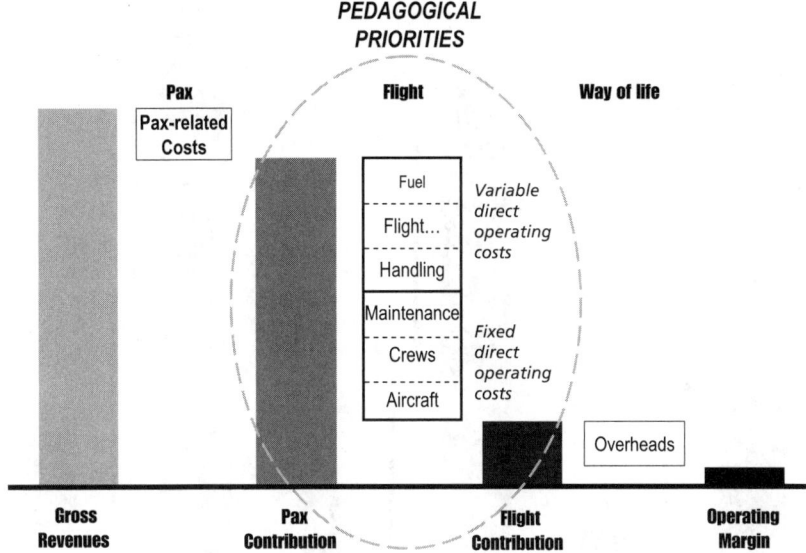

Figure 8.1 SN's business model.

taught us a management 'jewel', which is a must for anybody who wants to develop a general management attitude in their approach to any business challenge. The recurrent question that we heard at nearly all of his sessions was: 'How do we make money in this business?' Some readers might consider this observation naïve, but frankly it took me some years to really transform this question into a systematic reflex.

Xavier instilled in me this sense of constant research into the business model that is the expression of one's personal synthesis, which makes it possible to explain first to oneself and then to others how this business can thrive according to its specific constraints.

A simple test can tell you whether you have reached this synthesis: can you draw it? If you can, then it is possible to mobilize the company's business doers around the correct theme. By drawing the business model, I mean a simple 'revenue–profit waterfall', as in Figure 8.1 (pax represents passengers).

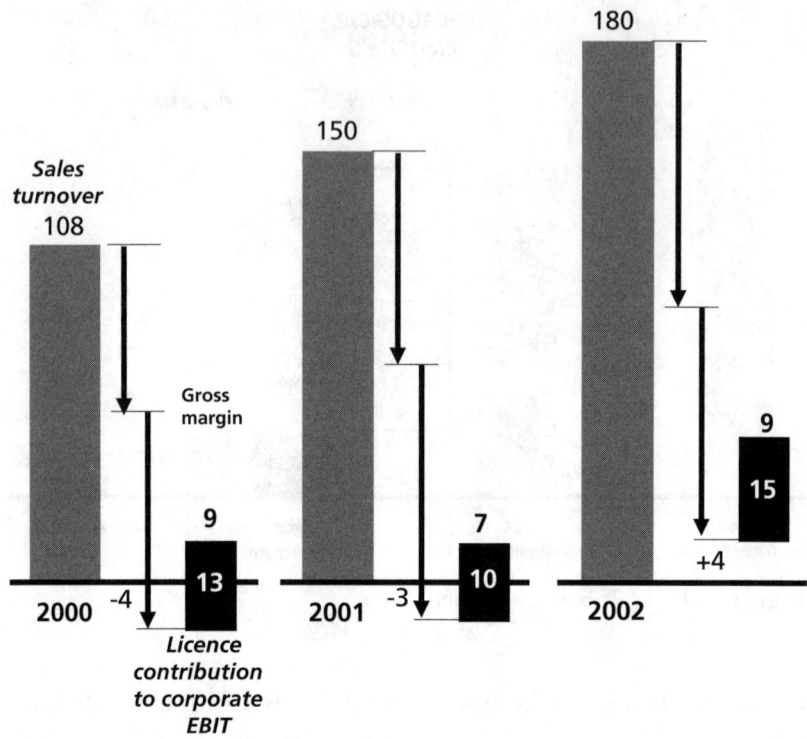

Figure 8.2 Kenzo's business model.

Attention is unambiguously focused on the central priority of SN's business model – direct operating costs – and consequently the whole analysis is a question of the break-even point at the elementary level of strategy execution. But when one observes the revenue–profit waterfall applying to the Kenzo case in 2002 (Figure 8.2), one can immediately note that the situation is radically different. Therefore research into the business model is a customized approach that cannot be taken for granted.

The lesson is to recognize this ad hoc way to make a profit. In the Kenzo case, observing the specific profile of this waterfall, what should the reaction have been when the CEO explained that the big strategic shift was to bring the licence business in-house? It sounds like the future profits were seriously in question.

Back to SN, its waterfall graph shows that the direct operating costs are the critical point to be observed to understand the performance dynamic, which must be able to explain the nuances of the level of performance reached. Sales revenue *minus* passenger-related costs represents the contribution of the route or a simple flight. Therefore the question is, is it sufficient to cover the direct operating costs (DoC)? Only two answers are possible, but they must be read with some caveats.

- *No.* So by flying an operational loss has been generated. But because the company is in a sector of heavy fixed assets, a caveat must be addressed. When one observes the structure of the DoC, one can note that part of them (aircraft, crew, a share of the maintenance) has to be paid even if one does not fly. These are really the 'fixed fixed costs'. To simplify, let's say that they represent in this case 40 per cent of the total DoC. Consequently, when one decides to fly the first goal is to cover the variable side of the DoC (fuel, handling flight-related costs – Eurocontrol fly-over costs, for instance). If not, this means that it was a bad decision to fly: on top of the fixed fixed costs some additional losses are incurred. In contrast, any euros on top of these variable DoC are a contribution to covering the fixed fixed costs. The reader can immediately smell a rat at marginal cost accounting where one can justify mediocre results just because they make some contribution that absorbs part of the fixed fixed costs. This way of thinking is unhealthy by nature, because it leads one to accept mediocre performance and to justify it in this way. Unfortunately, it institutionalizes mediocrity so that it becomes a routine. The former SABENA was a good example of this kind of practice. In the last years of its life, having acquired a number of Airbus 320s that greatly exceeded its real needs, the company started considering flying to some secondary destinations from Brussels with very limited market potential just because the assets were available on the tarmac. No need to explain that this accelerated its fatal decline.

- *Yes*. So we address another set of questions: is the contribution on top of the DoC sufficient to cover the overheads? If no, the company suffers a corporate loss. If yes, the flight contributes to corporate profit.

Drawing this business model leads to the next natural question: how are the performance leverages used to influence the result of a flight, or a route? One of the merits of the business model is to kill the fallacy of performance improvement to focus on the genuine means of achieving it.

Business story 8.1: Substitute Avro 100s with 85s

When a company possesses aircraft with different capacities, it can adjust its capacity to the legitimately accessible market. This option was not possible to SN, because the fleet of its European operations was initially made up of Avros with an average capacity of 88 seats, resulting from three sizes of aircraft, accommodating 82, 85 and 98 passengers. During performance review meetings, I saw a lot of time wasted on ping-pong discussions over the idea that a route was losing money because the correct aircraft were not allocated to it. I even remember a German representative from Hamburg presenting a performance improvement plan that included the statement, 'Substitute 98 seats by 82'. This unfortunately was not always possible.

But the truth lay somewhere else. One needs to go back to the central concern of the waterfall graph, DoC. The DoC per flight difference on the Hamburg route between an Avro 100 and 85 in 2003 was €82, representing the average contribution of one price-sensitive passenger. Therefore, in this specific context the choice of regional aircraft was clear in terms of performance. But between an Avro and an Airbus 319, the story was different: it was €1000 or an increase of 20 per cent of DoC, representing the contribution of 12 passengers per flight. So with 1800 flights a year, just to cover the substitution of the Avro by an Airbus the requested additional contribution was equivalent to 20 000 passengers a year, in a market growing only by 3 per cent with a total size of 130 000 passengers.

This example is just here to remind us that in a business with costly assets, perfectly matching the profile of asset deployment with the legitimate accessible market is crucial. But for the transformation process we can also observe that we now have available a simple drawing that will systematically put the finger where it hurts. So there is no way to lie to oneself about the effectiveness of the recovery.

APPROXIMATELY CORRECT NOW

Above I mentioned the importance of the basic toolbox dealing with physicals. The goal in SN was to broaden the audience of managers with whom to share this performance information and to give it the simplest possible expression.

I mentioned the significance of the waterfall as a means to express the critical issue of making a profit at the elementary level of strategy execution. DoC is crucial, but now the challenge was to shift from concept to management reality. The challenge was not to run the analysis itself but to create the conditions for a disruption in the way of dealing with this information. But a disruption on which theme? The challenge imagined in the turnaround of SN was: can we find a way to get the result of this business model in real time? Therefore the challenge was to instil a new corporate value in SN: 'Approximately correct now instead of accurately right too late.'

In SN Brussels Airlines' business, there is one piece of information that is nearly instantly known: the number of passengers per generic booking class on each flight. Broadly speaking, these can be called C for business-class traveller, TS for time-sensitive or full economic traveller, and PS for price-sensitive or budget traveller. The question is, with this information systematically available, can we assess the performance of each operation nearly in real time? Is it possible and with which level of accuracy?

In a turnaround context, we cannot rely on an advanced and sophisticated information system because we don't have the cash available to create it. Consequently, the challenge must be addressed differently, with a solid dose of innovation.

The challenge was first to introduce a performance measurement system that did not give the impression that the existing system was redundant. To avoid this trap, I had a powerful ally: the trick of measuring with physicals. So what were the physicals in this case? The physical in an airline is the passenger. Therefore I suggested that everything should be expressed in a new passenger-flavoured currency, which would help me because it would look like a trick and not threaten anything that already existed.

When SN was relaunched, the business context was highly influenced by the development of low-cost airlines, leading to passengers becoming more and more price sensitive. In the fare structure, more than 15 booking classes from the lowest possible fare to a very cheap one were dedicated to price-sensitive passengers. Then the same thing was done for time-sensitive passengers who were looking for more flexibility. The more they paid, the more flexible their ticket was. Finally, there were two or three business booking classes that do not need to be outlined here. In each segment it is possible to know the average contribution (revenue minus passenger-related costs) per revenue segment (C–TS–PS). So I proposed to express everything in terms of PS and

Value of traffic estimated in YEP

(2Cx4.1)+(5 TSx 2.7)+(66 PS x 1) = 88 YEP

Day	Flight	Aircraft	Flight BEP (YEP)	Perf. (YEP)	Flight Contrib. (YEP)	Overheads BEP (YEP)	Corporate Result (YEP)
....	....	....	....	....	....	....	....
....	....	....	....	....	....	....	....
Monday	3593	AR1	71	88	17	85	3
....	....	....	....	....	....	....	....
....	....	....	....	....	....	....	....

Each day

Figure 8.3   Example of YEP BEP on a specific route, Lyon, March 2003.

we introduced a new physical currency, YEP, an acronym for Y Equivalent Passenger.

For instance, in Figure 8.3 illustrating Brussels to Lyon with an average PS passenger contribution in the month of March 2003 of €76, representing 1 YEP, the TS average contribution was 2.7 YEP, the C average contribution was 4.1 YEP and the DoC per flight was 71 YEP. At this stage all the ingredients were in place to teach the concept of break-even point and get it assimilated. The next day for each flight management's question was, did we beat 71 YEP? Moreover, this could be known at 10 a.m. the day after. So with this physical approach a new mode of operational assessment was emerging with an approximation of performance nearly in real time. The figures we were using were not dummy ones. This was accurately validated in the system with very limited discrepancies, which were erased during this project because the estimation was done with more and more accurate data. At the end our estimates of contribution were based on what the passengers had generated only 20 days previously.

After the curiosity and interest of the first few days, we got the classical inertia champions with their bags full of objections or even scepticism. I heard an interesting comment from one manager: 'This YEP is the same as what we used for some performance measurements that we call full club equivalent in Swissair.' The former Swissair full club equivalent encompassed an endemic weakness. Given that there are fewer and fewer business–class passengers, would it be appropriate to take the risk of developing a corporate reflex with something that is not a reference rooted in current trends? YEP could find some analogy with the past so that it sounded almost familiar, but this physical was in line with the current trend of the business context. I did not ask for more from this manager: his comment was already a fantastic excuse to move on.

So a new corporate currency must reflect the trends of the current business reality. That is why, in the current context, the challenge must be addressed from the volume side, with price-sensitive passengers as a strategic point of reference. At this stage we considered that paradoxically it was with this population of customers that part of the future prosperity of the company would be recovered.

Three advantages are associated with the physical approach to performance measurement at SN:

- First, it broadens the management audience for the performance measurement system.
- Secondly, it develops the real-time tempo of performance assessment and consequently increases the speed of reaction. Each day, at 10 a.m., the estimate of the previous day's performance, correct 98 per cent of the time, is shared by 20 senior managers and all the country managers of the network. If it is available each day at 10 a.m., on the first day of the next month at 10 a.m. the consolidated performance of the portfolio of routes for the previous month is also available.
- Thirdly, it is difficult to copy.

This new corporate currency, YEP, is the visible part of an iceberg. It is surprising! You could hear during the pilot phase people asking, 'What is it?' But realistically, I must admit that YEP is nothing without its frame of reference, the route. It is the legend of a new corporate map, which one needs to be taught how to read. Consequently, investing time in SN Brussels Airlines' collaborators in order to allow them to read their own achievements from a different angle is a mandatory process before we can imagine them quitting their previous comfort zone. Finally, this new performance map must become a platform on which a new corporate dialogue can be developed, with new ambitions and a realistic sense of achieving some stretched goals.

The strategy execution monitoring system has found a brand new dynamic that can theoretically lead to the creation of a new mindset, where the priority is given to speed over exact accuracy and to speed of sharing information systematically presented in the same format. What has been described up to this point was an important breakthrough, but a more difficult challenge was in front of us: how could we make the management team fluent in YEP?

BEAR IN MIND

*I*n many companies over time, I have observed that they systematically develop some form of short cut in their performance assessment mechanism. The cause is simple: the growing size of performance information reports. These short cuts are performance measurement derivatives. In the airline sector rask and cask are characteristic of this situation; in a home decoration product distribution chain the derivative was more ad hoc, 'white paint' for instance. These tools were absolutely appropriate when they were first used, but they have either become threatened by a form of obsolescence that spoils their past relevance, or what they are reflecting is not appropriate enough for the current business challenge.

In a turnaround context, when one encounters these performance indicators, I suggest systematically avoiding them because they develop two crippling handicaps:

- First, they are an expression of a past that led the company to the crisis one is trying to tow it out of. So it is wise not to create confusion.
- Secondly, even if they are analytically and technically correct, they are often managerially inappropriate. This concerns not only how they trace a cause-and-effect relationship in performance measurement, but also how they enforce the appropriate corrective measures with the right tempo.

My objective being the management dimension of a turnaround, I therefore needed to find new ways to support the relationship of a just-reborn company to its true operating performance. This not only kept focusing the collective effort, but also reinforced confidence building by allowing management's attention to be focused on a positive message about operating performance.

Throughout a turnaround one of the goals is to empower the whole management team so that they have a healthy judgement of the situation for which they are responsible. To reach this stage two conditions must be fulfilled:

1. First, the senior management must create the correct context through its style to support a new relationship with performance measurement and also the necessary corrections. For instance systematically using them, organizing a formal presentation workshop on the purpose of the performance measurement tool, and finally enforcing a discipline of not using the old one any more.
2. Secondly, a radically different proposal for performance management information must be implemented. This combines:
   - The reintroduction of some basic management tools, which have been completely forgotten both in terms of relevance and use. They are very often BEP flavoured and based on physical elements instead of a financial perspective. This is frequently completed by the introduction of a new unit of

performance measurement. The purpose of this new unit is mainly pedagogical and consequently temporary, but it is mandatory to prepare on a solid basis the future return to a more financially flavoured measurement of performance.

- An insightful choice of the appropriate level of observation of performance. To do this, it is crucial to recognize, observe and monitor performance at the level of execution that is strategically relevant. For instance, in the case of SN in Europe this was the route, while in Africa it was the gateway; in a distribution sector, the shop; in the hotel industry, a specific hotel; in a decentralized service business like DHL, the station. Then any form of corporate consolidation can be build on top of that.

- A revisited business model, recognizing how the company intends to make a profit in its own specific context. This leads to a revalidation of the priorities on which management attention needs to focus.

- A performance measurement nearly available in real time, because this proposal accords with a fundamental differentiating characteristic: *approximately correct now rather than accurately right too late.*

This proposition is the formula for changing to a new relationship with performance. It confirms that the page has been well and truly turned from the past, and secondly that it is easily accessible by a larger number of people than previously. The clear evidence of this deep change is the emergence of a new corporate lexicon to deal with this topic, not only a new performance unit but also a new language.

PRACTICAL REFLEXES TO DEVELOP FOR JUMPING INTO THE HEART OF A TURNAROUND

- Do you have performance measurement derivatives in your organization? For how long have they been in place? Are they still relevant and why?
- Draw a P&L waterfall. Check the current hierarchy of performance observation: does it follow the same logic as the graph?
- Is the elementary level of performance monitoring in your organization strategically relevant? Why? Can the manager in charge explain how their own break-even point is calculated and which unit are they using to discuss the performance assessment with their staff and with their management?
- How is the performance observed in physical terms? Do you have a formal lexicon of your performance terms and how does it reflect the company's strategic orientation?
- How long does it take before you have a measure of yesterday's performance and the consolidated performance to date?

COMMENTS AND OBSERVATIONS

Pierre Henry, CEO, Sodexho Pass International[31]

*T*he heart of Jean-Frédéric Mognetti's observations concerns situations of deep, even fatal crisis. The vast majority of CEOs won't ever be confronted with a disaster of such seriousness, and that is why I interpret Kuijper–Davies' experience at SN as a very practical catalogue of techniques to be chosen and implemented in order to prevent such a situation from occurring. We can benefit from cherry picking among these concepts and this approach as appropriate.

A company such as mine – Sodexho Pass International, with 27 subsidiaries throughout the world supplying meal and service vouchers – operates in a portfolio management context, where almost half of the portfolio is continually performing below the average performance of the group. This is perfectly acceptable,

---

[31] See www.sodexho.com.

because it is a simple mathematical consequence that even those who like rankings must not forget. What is more annoying in our specific case is when former shining stars that lose their sense of direction cannot on their own find the resources to recover. It is in this respect that what Professor Mognetti describes is very appropriate in structuring a way out of the crisis.

For instance, in early 2000 the Mexico subsidiary was among the top three performers in the company. A recommendation from a leading consulting firm, which unfortunately was severely lacking in any understanding of business reality, suggested leveraging Sodexho's volume of business to increase the commission rate paid by our affiliates (mainly supermarket chains). This turned into a nightmare, but the company was dead to the repeated signals that the policy should be aborted. Three years later, the subsidiary was below break-even and all its efforts to get out of this tailspin were ineffective and frustrating. The damage resembled a bull let loose in a china shop. This was the situation when I took charge as CEO of SPI in 2004.

This example has a very important characteristic in common with the relaunch of SN Brussels Airlines: the availability of cash that we both enjoyed. This situation was radically different from a case where one must find a source of cash in savings that have not yet been achieved. Consequently, the main theme for the CEO to stress in such a situation is a radical disruption with the past, in order not to prolong the pain or allow any form of benchmarking. My recommendation is to turn the page as fast as possible, without wasting time either on the market side or within the organization. The mission of an orchestra conductor is to ensure that a musical dialogue takes place within the orchestra. To pursue the analogy, this team was only able to play notes, not music. So my objective was to enable them to play music again as fast as possible.

For this Mexican example, I will fine-tune Mognetti's proposal to get *back to basic tools* so that it becomes *back to basic tools through disruption.* My appointment as CEO was a trump card in the turnaround process, but it was one I could only play once. It was a very effective provider of leverage to turn the page from the past, with both customers and staff.

It is vital to get back to basic tools because sophisticated tools are useless in such a situation where a deep sense of the business has been lost. When one of our subsidiaries gets into a crisis it is the result of a loss of practical intimacy with the genuine meaning of the business model. I like Mognetti's expression that a business model expresses how we have decided to make money in our business. SPI's business model is twofold, drawing its revenues both from its value-added-based fees charged to companies for providing them with the tools supporting their fringe benefits policy, and from the commissions paid by affiliates (networks of shops and restaurants) for the business generated by those spending SPI vouchers. In Mexico, there was an attempt to enforce an ill-founded paradigm in respect of the affiliates.

From this experience I draw an important lesson: the business model is also a diagnostic tool. When the business model is fully actionable in the minds of the management team, it is also an aid to enable quick corrections to be made. In contrast, when this is not the case, the local team on the ground has lost its familiarity with the business model and is just paying lip service. We are in a situation of fake commitment. Moreover, when the micro-management is failing, macro factors often also add their own constraints to the company's performance. For instance in Mexico, the company was not only struggling with an inappropriate evolution of its commission policy with the affiliates, but simultaneously the interest rates in the country were falling. Since the company's performance is directly linked to the level of interest rates, this kind of event adds a second concern and makes the probability of getting out of this bad situation very low. So, *getting back to basic tools* is a management discipline that prevents a problem turning into a disaster.

In our case, in contrast to SN, we don't have to reinvent a business model because we are clear about that and its positive impact in other parts of the world. We have an easier task: we just have to enforce the correct, demonstrated practice. Only new management and a new structure can do this. The team who led the company into this crisis cannot tow it out of it. There is no time to waste. A new team must replace the old one; with the size of a group like ours we can always set up an interim team in order to start the turnaround process without delay.

I also made the company move to new premises. This move was not cosmetic or for cost reasons – it was a further means of reminding staff of the meaning of the business model. The former premises were not set up to cope with the two fundamentals of performance measurement in this business, which is driven by detail:

- 100 per cent on-time performance for voucher delivery. We are more punctual than any airline, our standard of performance is 100 per cent. If we do not achieve this it could mean a potential crisis for the customer, something we cannot afford or even imagine in terms of continually adding customer value.
- 100 per cent on-time performance for reimbursement of vouchers collected by affiliates.

These two indicators are paramount. If they are not respected, we spend our time running after the mistakes to correct them. I cannot imagine that in that context we could simultaneously be creative in developing new added-value solutions for customers.

The issue is to gauge how the way the company is physically laid out allows it to fulfil its tasks and cope with these two physical indicators of performance. After that we can always imagine a corporate dashboard. But if the fundamentals are not in place, we are putting the cart before the horse. Consequently, the move was a way to plug the business doers back into the reality of the business model. But as a CEO, I enjoy putting a cherry on top of the cake: the previous premises did not meet earthquake standards, so I was not unhappy to leave them.

Our business model is the frame of reference for our mission. The team in Mexico had lost confidence that they couldn't develop something relevant for the client. In refreshing their skills I consider that this team returned to superb performance. But in this approach we must not confuse haste with speed. I used the example of petrol vouchers to retrain the team in effective negotiation and incidentally rebuild their own confidence. Simultaneously, it was my responsibility to rewire the future relationship with the supermarket chains. If I had not used the petrol case as a means of recovery and instead pushed the team back in front of the supermarket chains, the whole company would have stayed stuck in a ditch, perpetuating the past. The middle management of this company did not believe in its potential for recovery. It was my mission find the silver lining in its culture to create a recovery dynamic, which found its energy in the depth of the company's pride. This is a game where we must be very careful in selecting an appropriate challenge.

What was done at this stage echoes what I can read in Part II when Professor Mognetti recommends taking care not to break the bounce of the organization by mobilizing it too early towards some difficult goals, but to redirect its limited energy in the correct direction through 'cluster breaker' exercises. That was exactly the purpose of the petrol example, when simultaneously it became my mission to rewire the business relationship with the supermarket chains to support the recovery tempo. What was great after a few weeks was that both the team and myself could share two successes with a solid synergistic effect to speed up the recovery. As CEO you have to pull the recovery manoeuvre to the forefront of the management system if your ambition is to attain sustainable results quickly, otherwise you can be lying to yourself.

PART IV

TAKING THE
CORPORATE PULSE

A t the end of Part III, I illustrated how managers can use a revitalized performance measurement toolbox as a key step for getting to grips with a turnaround. The process of revalidation was driven by a strong principle: to democratize the means of monitoring performance at the appropriate level of strategy execution, thanks to a clear business model and a unit for performance measurement – YEP – that is relevant to management and useful for informing the transformation process.

The basic approach of the toolbox is a never-ending iteration, relentlessly making the management technique as simple as possible in order to address so far untapped areas of management effectiveness. This fourth part of the book will dive deeper into the heart of the turnaround. With the help of some other examples, it will show how SN started on the correct foot and did not make any mistakes in insightfully matching the appropriate use of its asset base with its strategic ambitions and securing its positioning in the sector. Finally, pursuing the basic tools approach, the company fine-tuned the development of a unique performance assessment system or corporate pulse.® This continues to cement the confidence-building process started in 2002 due to its relevance, its speed of availability and its breadth of applicability.

CHAPTER 9

LEVERAGING THE VALUE
OF THE HERITAGE

THE ASSET BASE MUST BE
STRATEGICALLY APPROPRIATE

The basic tools described in Part III were already in place when, in mid-2003, they were embodied in a winning formula that made best use of the company's heritage. This was summarized in an appropriate asset base and combined with healthy, systematic strategic management and focused research into optimizing the business model. In my opinion, this winning formula represented the operational reasons for SN avoiding the trap into which Swiss International Airlines fell. This chapter details the three components of the formula in order to assess its management value and the lessons that can be drawn from it.

In Part III, I underlined that one of the fundamental characteristics of operating in a turnaround context is the need to rebuild internal confidence; especially on the part of individuals, who provide the cement for the collective confidence. This is always an ambitious goal, which can quickly become no more than a dream if the strategic conditions are not favourable. In this respect, SN was lucky. For the relaunch of SN Brussels Airlines, the new company's asset base, its fleet of Avro regional jets, represented a favourable opportunity to fulfil the new company's mission, 'Connecting Brussels'. This fleet, with a range of from 85 to 98 seats, corresponded to Brussels Airport's average number of passengers per flight (see Figure 5.1). Consequently, the asset base was structurally well suited to match the ambition to regain traffic in Europe.

It is a prerequisite in a turnaround to assess objectively whether or not the company's asset base allows it to join the race. It is an area where a lack of insight can be very costly, even fatal. In this respect, one must not mistake the trees for the forest. It is not only a question of having the latest equipment but also how that equipment is used.

In the 1990s I advised a private equity fund that specialized in taking control of bankrupt companies at the courtroom door (see Business story 9.1). Of course there are some spectacular examples of success doing this, but the vast majority are failures, because the company's asset base cannot match the evolution of the competitive context without a massive investment in refurbishment, which an acquirer is not always ready to make.

Business story 9.1: Same assets, used differently

One important question consists in assessing how close the company is from its sector norm in terms of assets. The private equity fund I was advising – Altus Finances, part of Credit Lyonnais – was buying companies that represented 'le prestige de la France'. These companies had usually been through some management turmoil and needed to be repositioned to get back to a place in the market that they should never have relinquished.

My collaboration concerned a specialist pantyhose brand, Gerbe (www.gerbe.com). Besides its own brand management, Gerbe also manufactured for Dior, Sonia Rykiel and so on. When I visited the manufacturing premises of this company based in Burgundy, the CEO explained to me that 60 per cent of its pantyhose-knitting machines were brand new and could easily rival those of the world leader in this upper-end segment, the Austrian brand Wolford (www.wolford.com).

I bought the argument, but some weeks later in this project visiting the premises of the Austrian competitor, I discovered that the challenge in terms of assets was not the manufacturing equipment, but time to market and how the assets allowed the company to succeed in that. So, the question was how many market tests had to be made to gauge the potential value of a product. Wolford owned a dedicated pantyhose prototype mill, while Gerbe made its prototypes on its normal manufacturing equipment. The CEO had visited Wolford, but the company did not show him this part of its process, because it was exactly here that some dedicated assets were making a difference and creating a competitive advantage. Wolford was not more creative than Gerbe, but it had the means to test three to four times more prototypes. It was more often at bat, as one says in cricket. So the odds were in favour of Wolford. The reasons for its success were its manufacturing organizational skills and a focus on time to market achieved by a specific way of managing its assets.

In summary, the assets side of a turnaround is crucial. This issue is very high on the agenda when the company's challenge is to renew its assets, because there is a need to assess whether the business model has the capacity to support such an investment. This question needs to be at the back of the CEO's mind very early in the turnaround process. If not, the turnaround runs the risk of being no more than a remission for a limited period of time. However, for SN asset renewal was not an urgent topic. Consequently, it did not have a handicap in this area, which was already a great advantage.

SN Brussels Airlines was created with the correct asset base to connect Brussels. But objectively this merely represents a 'wild card' for entering a grand slam tournament, nothing more. To support this statement, let's note that Swiss, born almost simultaneously, operated with a radically different asset base. Its portfolio of assets was far more prestigious than SN's, but less than 18 months after its relaunch Swiss was already sending back some of its brand new Airbus 330–340s to the leaser. Consequently, being created with the correct asset base is meaningless if one does not match this opportunity to one's legitimate and accessible ambitions; although it is also fair to say that one's ambitions must not be beyond the range of one's assets. SN successfully passed this hurdle of 'fitness' in remaining consistent with its 'Connect Brussels' ambition. In Europe the objective was to revitalize the healthiest part of the Sabena network; in Africa the goal was the same, including some opportunistic connections in West Africa that had attractive business potential due to the possibility of an exclusive service.

This test of the ambitions–asset match is never a waste of time. It avoids one being confused about the league that is from now on the most appropriate. In this respect, SN was sending an additional encouraging signal that it was not repeating its past, while Swiss could not free itself from its tailspin pattern.

THE COMPETITIVE SITUATION OF THE ROUTE DEFINES THE STRATEGIC GROUPS

SN's European portfolio of connections incorporates a broad diversity of destinations, including country capitals, regional capi-

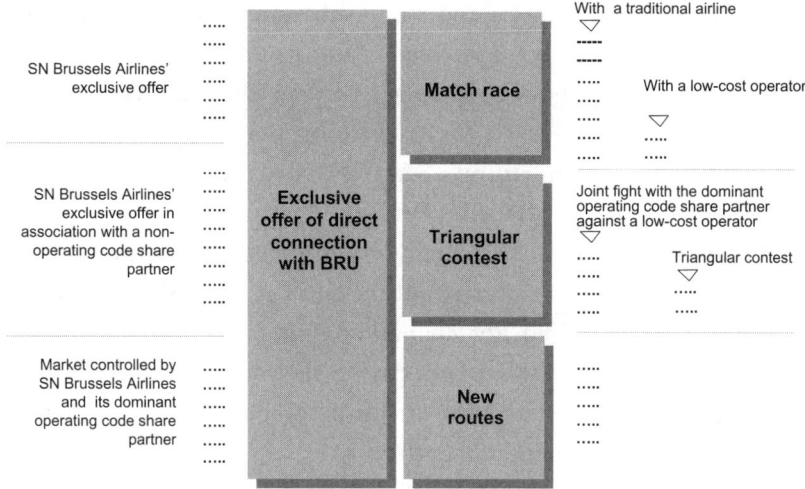

Figure 9.1   SN's European strategic group of destinations.

tals, leisure and ethnic destinations. Addressing this portfolio of destinations according to its competitive conditions led to its being resegmented into three strategic groups: Exclusive Routes, Match Race Routes and Triangular Contests. A fourth category – Incubator, for new routes – was added at the end of 2003. The rationale behind the approach was the radical difference in the competitive challenge for each group (see Figure 9.1).

- *Exclusivity* concerns situations where SN is the only operator to offer the connection between Brussels and the destination airport. This category is subject to additional subsegmentation with respect to the respective code share conditions.
- *Match Race* is where SN is in direct competition on a specific destination, either with a traditional airline or a low-cost operator.
- *Triangular Contest* is where more than two operators offer a service between Brussels and the destination airport. This usually involves a traditional airline and a low-cost carrier. Virgin fulfilled the low-cost role in many cases. SN's acquisi-

tion of Virgin Express and some subsequent withdrawals have not fundamentally improved the quality of performance, because Virgin has been replaced by another low-cost operator. This was observed for instance on the Copenhagen route, where the competition SAS–SN Brussels Airlines–Virgin Express became SAS–SN Brussels Airlines–Maerks Air. The margin pressure is structurally conditioned by the number of players and their respective profiles. In the case of Copenhagen the substitution did not change anything.

- *Incubator or new routes* regroups the development situations of routes initially tested with off-peak capacity and consequently targets a less demanding performance.

From the beginning, it was obvious that Exclusivity routes were naturally generating a better level of performance than the other forms of competition. Match Race offered modest and more laborious returns on effort, although nonetheless encouraging, while Triangular Contest often turned into a blood bath.

The attractiveness of Exclusivity is due to the absence of over-bidding in competitive pressure, which almost systematically develops as soon as two operators are neck and neck in competition. The airline is a business where any newcomer on a route always considers that it can gain a substantial market share because it will compete on price against an incumbent. This just creates a temporary pressure on margin, which will more or less reflect the relatively relentless style of the incumbent. The latter can play with a wide range of weapons, even sometimes borderline ones. But these are the rules of this game, which is cut-throat not gentle. For instance, this includes pressure on slots in order to prevent the newcomer matching the offer in terms of service (frequency and schedule), structurally forcing it to face a severe handicap.

This is not exclusive to the airline sector. I had the opportunity to observe the same kind of situation in the cellular phone industry in a form of 'management laboratory', the islands of Trinidad and

Tobago. Usually, when a newcomer attempts to break a monopoly the result after two years, due to its efforts to buy market share, is a reshuffling of market share to 60 per cent for the incumbent and 40 per cent for the newcomer. In this case it was just 80–20. The explanation was simple. First, the newcomer was not outstanding in its launch. Secondly, the former monopoly transformed into a stronghold its access to the more lucrative segment of the contract market; fought with a systematic, matched approach the prepaid volume market; and finally did not exactly share its network in the way imagined by the local telecom regulator. This could be judged as borderline and needed to be settled in court, but this was too late in terms of market perception.

This case from the West Indies had similar characteristics to the fight between Air France and AirLib on Marseille–Paris, where Air France forced its competitor to satisfy itself with losing money on the very low-contribution fare. Even though Air Lib enjoyed an 85 per cent seat load, it pulled out after two years. The situation was not tenable for very long.

This is what faced SN in its competition for exclusivity. It even led to the ability to charge a relative premium for its services. This has two main origins: the absence of competition resulting from the other party's withdrawal, for instance on Bologna once Air Dolomiti stopped its operations; and exclusivity resulting from a joint rationalization of asset deployment thanks to a partnership agreement, for instance on Prague. SN then benefited from operating costs that more closely matched its legitimate share of revenue. This group of exclusive routes was driven by one goal, to make an operating profit (which we named a fleet contribution in SN's lexicon) or to diminish the losses. This led to efforts being focused on leveraging the appropriate means for each specific context. Consequently, the negotiation of code share agreements became a strategic axis to secure the relaunch by diminishing the costs of exposure to fiercer and unnecessary competition.

AD HOC JOINT-VENTURE BUSINESS MODEL BOOSTER

Over my 40 months at SN, the code share policy was implemented or was on the verge of being so on the whole portfolio of routes, except in cases of competition with Air France and nearly all air companies outside the STAR Alliance. I am very positive about this policy, because it has helped optimize SN's resources and prevented unnecessary waste for so-called strategic reasons.

First of all, this strategically flavoured code share practice deserves two observations:

- It is not a panacea but a necessary evil. It releases the competitive pressure, which is critical for a just-reborn company. But simultaneously, in a context of market-limited growth, it quickly leads to a ceiling in terms of performance improvement potential. Consequently, it must be driven by the objective to prevent waste due to unrealistic ambitions.
- The absence of head-on competition must not lead us not to recognize a form of hidden or insidious competition, such as the 'metal only' rule that is more and more being observed.[32] But again this attempt, beyond being a classic characteristic of a partnership, reflects a genuine characteristic of an unbalanced alliance. To avoid this trap there is only one answer: scrupu-

[32] With a 'metal only' restriction, the customer of the airline company which is operating on code share on a destination is offering special conditions to some specific customers if when flying on this route, they book on flights operated with the company's aircraft (metal). In a code share agreement of free-flow type the partners keep the revenue of the passengers transported on their metal whoever issued the ticket. Therefore in order that the code share works well, this supposes that both partners are equivalent in terms of calibre, volume and metal.

lously manage the P&L of the partnership to truly assess it from the perspective of relative contributions. SN was able to do this, while doing so was way down on the priority list of many of its partners. Once again, we can observe that those who have correct, unquestionable information influence the discussion and avoid waste of time and frustration. Finally, it is the responsibility of the weak, SN in this case, to be overinformed to anticipate potential frustrations.

> In a partnership, it is the responsibility of the weak to be overinformed.

Code sharing remains a powerful means of improving the quality of performance, but it requires the company to stay permanently on alert against a softer but more perverse form of competition. Code share management is also 'coopetition'. But without code share agreements the company would not have reached its current level of performance in Europe. Consequently, I was strongly in favour of code share agreements and especially SN's specific management approach, strictly monitoring precisely what each partner is drawing from this close collaboration.

Code sharing is also an area that delivers some lessons in terms of strategic positioning and ambitions. In December 2005, while I was in discussion with Patrick Alexandre, Directeur Général Adjoint of Air France, an Airbus 320 was taxiing by his office, which has a direct view onto the tarmac of Charles de Gaulle Airport. He said:

> For five years Air France metal has not been flying any longer to Helsinki, it is Finnair that conducts this service. There is no ambiguity about code sharing at Air France, we know what is strategic and where our metal must be present. For the other destinations, our partners are a critical part of our strategic system. Take the example of Spain: we don't fly any more to the regional destinations

of Spain with our metal, Air Europa is running the service while we are fiercely competing on Madrid.

It is interesting to read this comment and to compare it with an observation in an article in *La Libre Belgique* on 3 January 2006: 'developing code share agreements with a marketing carrier status has contributed to weaken the commercial position of SN on these routes.'

This comment reflects a lack of knowledge of the nuances of the business model. Thanks to code share agreements, SN was able to increase its network at nearly no risk. Without such agreements SN would not have been able to do anything operationally on this route, since the potential losses would have been detrimental. Consequently, SN would have left the exclusivity of the operations to another carrier that was glad to find SN's commercial support in this agreement to help use its metal.

Or consider this answer to one of my e-mails in September 2006:

Jean-Frédéric

Thanks for the provocative paper, reading through the paper there is probably one thing where we completely differ from opinion, code sharing is intrinsically competitive, it is a means and not an objective. I consider it as an act of weakness . . .

Paradoxically this comment occurred at around the same time as the departure of the CEO and Executive Chairman from SN. It meant, in my opinion, that the company had lost part of its strategic common sense. With this kind of comment, some SN senior managers sent a signal of behaving like a major carrier, which SN was not. This point is critical for me and I interpret it as a loss of focus, the company forgetting what it was. This translates into not giving enough time or support to a measure for deploying its real strategic impact. On the routes to the big European capitals, either SN did not operate with its planes, or when it did so it was just as a follower, because there was always a leading major carrier and a low-cost airline in the vicinity. Consequently, the challenge was to understand under which market conditions SN had attained the status of first-league player. This was very clearly to the European regional destinations such as Turin or Bologna, which could be complemented by some ideal situations for strategic rationalization.

For instance in the UK, BA exclusively operated with its planes to London Heathrow, while SN had a reciprocal exclusivity with its planes on routes between Brussels and Bristol–Manchester (50/50) or Birmingham–Newcastle. This kind of example is called an 'alliance of complementary equals'.[33] Why was this agreement between the no. 3 airline in the world and SN a complementary alliance of equals? Because the strategic roles were clear. BA found an answer for supplying its hub at London Heathrow and SN became a connecting platform to reach either Brussels or other European destinations from the British provinces. Consequently, when the strategic roles are clear and when the assets deployed by both companies remain balanced, the alliance sticks to the above situation. But when one of the partners dominates the route with its planes, it is a bootstrap alliance, because the weaker expects to use the alliance to improve its capabilities. Unfortunately, the weaker partner remains weak.

> **A partnership is never a means to offset a strategic handicap; on the contrary it makes it worse.**

This is what happened with SN to Helsinki with Finnair, at least until the end of 2005. Since 2003 SN and Finnair had been operating in code share to Helsinki. The route was losing money from the beginning. The direct operating costs were the

[33] Joel A. Bleeke and David Ernst (1995) 'Is your strategic alliance really a sale?' *Harvard Business Review*, Jan.

highest in Europe due to the distance. Finnair was offering flights that were twice as frequent as SN's. Consequently, the higher-contribution business traveller traffic was not flying on SN. With this agreement SN expected to reach break-even quickly, but life is not that simple. The agreement just signed, a newcomer arrived on the route when SN was beginning its recovery. This new-comer was SAS's Trojan horse, Blue One. Its competitive goal was not SN but Finnair. As they say in Africa, when the elephants fight the ants take the beatings. That is exactly what happened. SN's performance relapsed into deep red ink, while Finnair was relatively protected. Finally, 15 months after the launch of this attempt, Blue One gave up operating between Helsinki and Brussels. Good news for SN! The Finland country manager was expecting a quick return to break-even, which did not materialize. The reason was simple: beyond the not so cooperative attitude of Finnair to rebalance the agreement, the market did not vote massively for SN, whose offer was not on a par with Finnair's. Consequently, when two companies are operating on the same route with a code share agreement the lesson is always the same: the offer must remain fairly well balanced in terms of frequency and product characteristics (schedule–fare structure–aircraft comfort) that offers passengers an equal level of flexibility.

An imperfect deal, consciously recognized as such, is always better than the heavy losses associated with a marginal role on a route with a dominant player, where your future does not depend on your commercial skills but on choices imposed by the leader. Consequently, when a company is a second-tier player it is wise to recognize that a substantial part of your destiny is in somebody's else hands. Code sharing is an asset-optimization tool, not a means to erase a handicap. It is a situation of shared pain and gain influenced by the fair balance of the investment and the relative quality of parity of the market offer by the two partners. The perceived fairness in this context is influenced by the quality of the information system. SN's system was outstanding, and it helped assess the real value of this strategic approach. Whether this situation can be transferred into other sectors is a never-ending debate; for instance, the situation between in-house distribution channels and independent distributors is of exactly the same nature as the code share agreement. It is pretentious to announce that one has found the correct answer in this area. It depends on the context, but it is easy to fool oneself if the correct performance measurement system is not in place. It was the same for SN. In its context the code share agreement was a great performance booster thanks to superior skills in performance assessment. This gave the company the lucidity to read objectively the benefits and simultaneously the limits of such agreements.

Consequently, SN developed a winning formula that optimized its asset base. The strategic groups were redefined by competitive challenge and the company was wisely paying attention to not wasting its resources thanks to its code share policy. One of the positive lessons of the code share approach was to say that SN was sharper at reading its performance than its partners. Consequently it could assess more objectively the impact of its policy. In Part III we discussed the principles of performance measurement, an approach mainly influenced by the use of some basic management tools such as BEP analysis, with a result expressed nearly in real time in a physical unit instead of a financial one. This represented a management proposal, which must now pass the bar of implementation in the company's day-to-day life to claim a contribution towards progress in terms of management efficiency. Having just addressed how the strategic frame was fine-tuned to deliver more management effectiveness, now I propose to describe the details of implementing the transformation process in terms of performance measurement.

CHAPTER 10

REWIRING THE PERFORMANCE ASSESSMENT SYSTEM

FROM SEAT TO FLIGHT

Trying to find the meaning of performance is a question of method. The rule is simple and my goal was to focus a large number of people around this method. Karam Kashani, one of the leading marketing authorities at IMD in Lausanne, launched me on a professional career with this simple but so useful rule: 'Massage the figures and they will tell you something.' This means don't stop at the result, give it a management relevance that can organize the appropriate reaction on a large scale. When I began to become more familiar with SN at the end of 2002, I realized that the company didn't have the tools to move in this direction. I therefore had to discover them.

In such a challenge, a classic trap is to jump on the existing performance derivatives and not validate their management relevance with respect to the context. At SN, as we have already seen, rask and cask were absolutely correct in terms of analysis, but I quickly assessed them as not meaningful enough to answer the 'empowerment and accountability' challenge, which many executives were embracing in their respective area of responsibility.

In SN's case, operating by analogy with similar situations experienced in the express shipment industry and in retail, I logically came to the conclusion that the performance measurement focus in SN needed to shift from *seat* to *flight*. In express shipment many of the performance ratios were parcel or document related, which was technically correct, but managerially speaking at the elementary level of strategy execution, the station, this was not of the highest managerial relevance. An observation per shipment would have been more helpful. It was the same situation in retail, observing performance by square metre while the empowered-accountable manager was managing a shop, so the analytical tool was not providing him or her with the most appropriate clues for management. At SN the objective was to support in the most effective way the relevant diagnosis and appropriate reaction of those accountable for the performance at the elementary level of strategy execution.

The rationale behind this decision to shift the focus from seat to flight was twofold:

- The level of performance assessment must reflect the practical reality of how the costs are incurred and contribute to generating some added value. It was clear that this took place at the occasion of a flight, which incidentally is made up of seats. Therefore the seat focus is contaminated by a form of over-analysis, which does not help with the development of a sound management attitude based on prioritization and synthesis.
- With an operating cost structure made up of almost 60 per cent variable costs, I consider it very risky not to put this dimension of the business under the spotlight to continually have in mind whether it was a good decision or not to fly.

This needs to be followed by the reflex of whether this specific flight is breaking even or not.

The correct focus is one thing. But it must also quickly become tangible for a critical mass of people to overcome the traditional inertia of any organization. This is not natural in a bankrupt and just-relaunched company, because the staff stick very firmly to what can be a reference point. So the challenge was greater because the ambition was also to change the reference point.

Paradoxically, the post-SABENA bankruptcy context represented a silver lining in this transformation. Joël, SN's database manager with whom I worked very closely at the beginning of this performance measurement reengineering project, made the following comment while we were celebrating with all the stakeholders the end of our first run of RBS meetings (see Part V) in early 2004:

> It is the first time I can see all the executive hierarchy observing operational performance through the same indicators with the same meaning for each of them. You couldn't have done that in the old SABENA nor at Swissair . . . the weight of the routine and the arrogance would have been too strong to imagine that the performance could be read from a different angle of strategic relevance, which they were already doing routinely. It is not a question that they won't follow you intellectually, but of paternity.

The earthquake of a bankruptcy was necessary to free this organization from its dramatic not-invented-here syndrome, which now allowed us to observe the performance of a flight at the route level. The result was presented as a break-even analysis expressed in YEP or Y Equivalent Passengers, as explained in Chapter 8. But this step was quickly shown to be not disruptive enough to plug the company into a radically different relationship to its performance measurement approach. Therefore we had to become even more innovative in the transformation of the performance assessment mechanism.

FROM YEP TO RODOC

YEP, the nickname of the new performance measurement approach, was launched in February 2003 for the portfolio of European routes. YEP quickly became a reference in the corporate lexicon – 'What has happened this morning? The YEP is not yet on our screen, it is already noon!' – as a new way to look at performance measurement. But six months after its launch, YEP was already entering a new stage of evolution, using this performance measurement technique in a way that would transform management attitudes more deeply. The idea was to use a performance measurement approach that reflected the driving costs of the operation at the route level. Based on some other experiences, I also had at the back of my mind the goal to complement it as soon as possible with a financially flavoured indicator that could also reflect the organizational effectiveness of the company.

Once found, the answer seemed pretty obvious: RoDoC or return on direct operating costs perfectly fitted this objective. I found it playing with Figure 10.1, which was a reference docu-

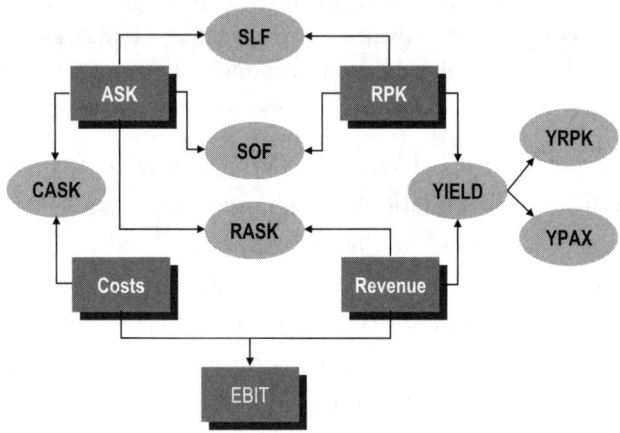

SLF: Seat LoadFactor, capacity/# of pax; CASK: Cost per Available Seat Kilometre; RASK: Revenue per Available Seat Kilometre; RPK: Revenue PerKilometre.

Figure 10.1   Traditional revenue–costs relationship in the airline sector.

ment in SN. This document was absolutely correct analytically speaking, but it left me frustrated. Under what is now DoC one could read in the past Costs; and under Pax Contribution, Revenue. In July 2003 I discovered the cause of my unhappiness with this document. It was too generic, even though it was correct! It didn't reflect the heart of the business model. What was critical was not the costs but specifically the direct operating costs, not the revenue but the total pax (passenger) contribution of a flight.

Direct operating costs were crucial and amounted to around 70 per cent of total costs. These specific costs, by their nature reflecting the *raison d'être* of the company, represented a mandatory reflex point in a management philosophy focusing on what is essential. It is fair to assess this cost accounting proposal as a form of 'back to basics' approach. In early 2007, Michael Tracy,[34] in a conference to which we were both invited by TSTT,[35] reminded the audience that cost accounting is an area that in terms of innovation is 100 years behind and often prevents us from looking at growth from a different angle. What he said is partly explained by Joël's observation a few lines above. This was exactly the approach followed at SN: break the old paradigm so that the organization can think freely and execute the performance recovery differently.

The pax contribution was also driven by the same straightforward approach, to focus on what one could rely on. The pax contribution is what the company can count on to cover its direct operating costs, because the gross revenue or total yield is

---

[34] Author of *Double Digit Growth*, London: Penguin, 2003.
[35] TSTT, Telecommunication Services Trinidad and Tobago, the former national telecom monopoly of the Republic of Trinidad of Tobago, currently owned 51 per cent by the government and 49 per cent by Cable & Wireless.

automatically reduced by the passenger-related costs due to their presence on the flight. Because a passenger is there the company has to pay the costs of acquiring that passenger. This can sound a bit didactic, but it is not as obvious as it appears to get it assimilated into day-to-day business life. A lot of saliva needs to be wasted to reach the necessary level of transparency on such a theme. In presenting these concepts to the different departments, the part of the approach that had the most impact was when I was explaining it via the analogy of gross salary vs disposable income. The company should rely on what is available to pay its core expenses, those related to its mission. That was crucial. But simultaneously, it did not mean that the company would not constantly look at increasing the pax contribution thanks to optimization of the pax-related costs. But this was just mandatory fine-tuning; it was not related to the structure of how the business model operates.

Therefore, the concept of RoDoC, which was just common sense, became a central part of the performance assessment system. This evolution occurred quite naturally, like a missing link in the system, the other ratios of which were already quite well known. Therefore I hid a substantial development under the guise of an add-on. Always obsessed by ease of implementation, this situation was for me a guarantee that the development would be accepted, and therefore its contribution as a key brick in the corporate confidence-building system was virtually assured.

RoDoC was also a clear graphical translation of the business model. Consequently, when we began to shift the focus from YEP to RoDoC this did not sound like a new fad, but more like an evolution. Therefore the risk of confusing managers remained under control (see Figures 10.2 to 10.4).

Finally, this performance measurement system was planned to become highly transparent at the corporate level. On a plasma screen in the company lobby, any staff member could see the daily updated RoDoC situation, which permanently posted the

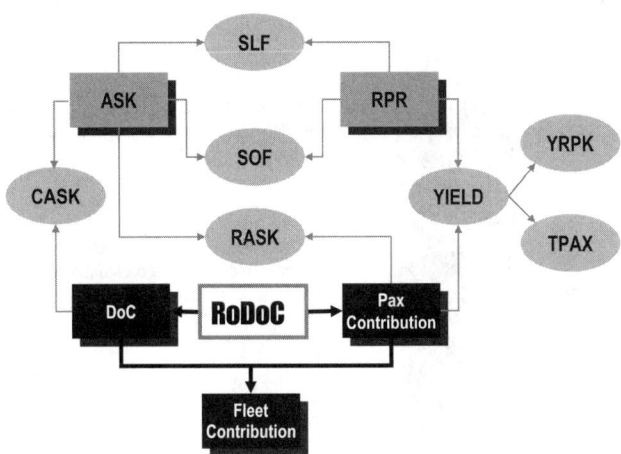

Figure 10.2 From traditional revenue costs relationship to RoDoC.

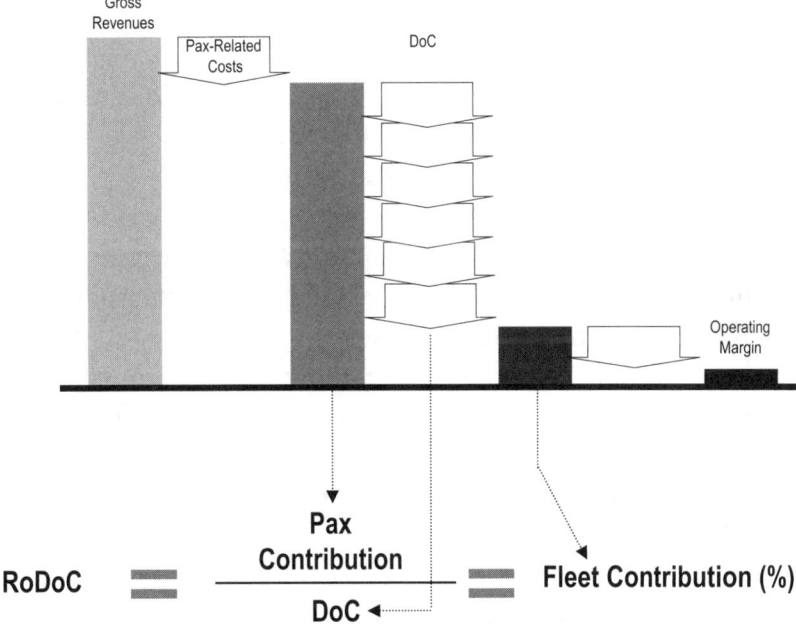

Figure 10.3 RoDoC and business model (1).

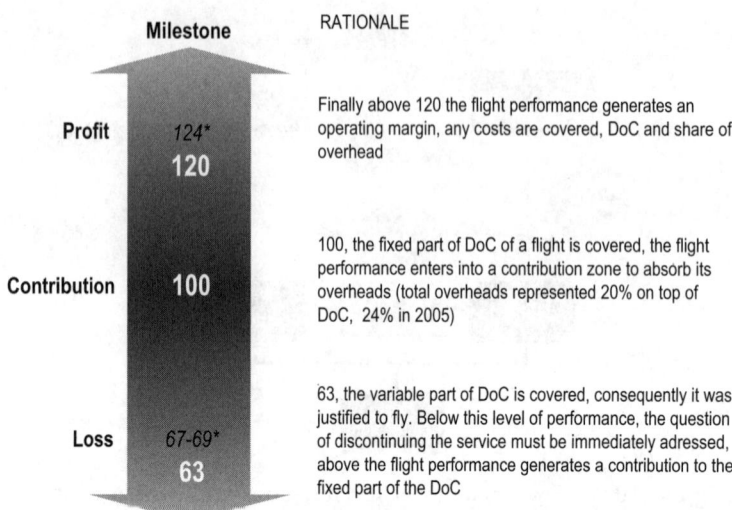

Figure 10.4 RoDoC and business model (2).

performance of the previous day and the consolidated situation to date (see Figure 10.5 later in the chapter).

In terms of development RoDoC represented:

- A major breakthrough over confusion about the nature of costs that could be observed at the relaunch of the company.
- A scale of performance achievement. With RoDoC, there is only one way to think: a flight either breaks even or it does not. If it performs under BEP, the question is by how many points of RoDoC in order to assess whether it was a good idea or not to fly. If it exceeds BEP, then the question is whether this contribution is sufficient to support the company's standard of living and the necessary resources to prepare for the company's future, such as fleet renewal. Finally, all these cases are associated with a colour code, which speeds up a common corporate diagnosis. This just represents a useful academic 'trick' to prioritize reactions.

RoDoC was intended to become a corporate pulse, not a simple performance driver. A driver is a business condition that when manipulated or changed will directly and predictably affect performance (see deviation analysis and for instance how performance was affected by capacity management; Figure 12.1 page 222). The corporate pulse® reveals the fundamental relationship between the company's main business drivers, in order to observe performance improvement and operational effectiveness.

The importance of RoDoC was first its impact in clarifying cost accounting. This exercise is relevant for any sector of industry. Behind the classical waterfall graph (Figure 8.1 page 135), one can find the key to many concerns. For instance, in supplier–sourcer relationships the classic and not so systematically achieved pocket-money analysis is a key analysis to understand what has been relatively given away to a customer and to gain which result. But the first key question is to trace this pocket-money waterfall to understand its dynamic and share it with the entire corporate staff. A similar approach was followed at SN when we introduced the concept of passenger-related costs. The challenge was the same: do people understand what it is all about? The answer was unquestionably yes.

In the specific case of SN the pocket-money side concerned only four types of costs: sales and commission, booking fees, passenger insurance and catering. Was this new? Not at all. The impact came through associating these costs into a homogenous bundle, which had never been done before. In this way I was able to address one critical challenge: getting employees to accept that what they've always done may not be the best way to look at reality. By this simple means, reality rushed back into their daily business life because the analogy with their personal life started to be clear.

An individual earns a gross salary, with social taxes representing the difference between the gross and the net. This net income to some extent is what is available to contribute to family development. It's the same at SN. Obvious, certainly, but I have also been an academic for more than 25 years and I am quite accustomed to see how faces change when an audience gets the feeling that what is being said clicks and that we can move to the next step. That is why I am still dealing with what to many others will appear to be obvious.

In summary, passenger-related costs represent money that does not belong to the company and it is not a priority in building and sharing a business model to try to imagine how this money can be optimized. I have just said that it is not a priority: there is certainly a lot to gain from booking fees, but this step was too far away for me at this stage of the transformation process.

RoDoC at SN or working capital at Nokia or net portfolio growth at Rieter are financially flavoured indicators, the first goal of which is to measure the operational effectiveness in their respective organizations. But these corporate pulses are not only financial in nature. In some other consulting assignments I came up with more functional suggestions. The nature of a corporate pulse makes it focus on what it is critical to monitor in the business model in order to start positively affecting performance and growth.

Business story 10.1: All the outlet sales advisers also become controllers

In an example in the retail sector with house decoration products, the core business of which was tailor-made curtains from an in-house fabric collection, the corporate pulse was the number of curtains per salesperson per outlet, with an elementary level of strategy execution monitoring at the shop level. Bear in mind the observation of Arnaud Fayet about the number of corporate pulses (see page 23):

I am happy with one even though it is a bit limited, but this number cannot exceed three for a simple intellectual constraint. It is not possible to get all the details in one's mind when one multiplies the number of indicators. Consequently, you lose sharpness of analysis and the conclusions are not as salient.

When we decided to add a new layer of corporate pulse, the commercial director and the shop directors thought that the price of the fabric was an important piece of information to trace. This was of minor interest in terms of value of execution. The shops that outperformed their colleagues achieved that not because they were able to sell more expensive fabric, but for a more fundamental reason. They sold more from the accessories product lines. Consequently, the next layer of corporate pulse was the number of accessories transactions per 10 curtain transactions.

This sense of measurement of the physical is crucial. And it represents a differentiator in terms of operational excellence, which needs to be continually reviewed in order to avoid backsliding. In term of follow-up no sophisticated system was necessary in this context, just a simple sheet of paper per day by the cashier. After each sale the saleperson just ticks with colour pencils the boxes corresponding to the structure of the transaction (the sales team usually consisted of around five people). The follow-up was easy, as was the analysis.

For many years, I have shared this obsession with measuring the physical with some of my former students, who have become friends and alter egos in this way of thinking about business performance. One of them, Frederic − who went to IMD after his first years of business experience in the French Mecca of luxury goods, Louis Vuitton − spent a few years teaching part-time with me. But for a couple of years he was also the retail marketing manager for the Swiss leather goods manufacturer Bally. Whatever the collection in the New York flagship outlet, his message was the same: so many pairs of shoes per day per customer adviser and so many bags. When he joined Nespresso, he was pleased to discover that it was measuring the performance of its shops in terms of number of coffee capsules and coffee machines sold. He shared with his new colleagues his satisfaction at seeing that such fundamentals were already in place. But he was more disappointed to note that for the new line of products, the accessories (coffee cup, tray, chocolates), the performance was expressed in euro sales turnover. Two physical indicators and a financial one, this was not correct. He immediately transformed the euro target into a number of transactions to be achieved per thousand capsules. All the outlet collaborators became controllers.

A second example of a transactionally flavoured corporate pulse can be observed in proposal-based transactions for companies operating a go-to-market business model. This situation can be divided into three main categories:

- The simple one, a form of transactional business where the goal is not even to remind the company of the customer's needs, but more simply to say what exactly is the supplier's offer at what price with which delivery constraints.
- The second category is more consultative selling, requiring some form of customization.
- The third category is where a truly new solution must be developed for the customer and both parties at the end of the day are mutually dependable.

In the go-to-market businesses in the second and especially the third category, I came very early to the conclusion that the difference between superior-performing companies and average companies is their level of dead proposals. Like pocket-money analysis, proposals follow a waterfall analysis, which first has to be clearly recognized as a reference point. This takes me back to the electrical installation industry in one subsidiary, based in Toulouse, of what would become the energy division of the Vinci group. Fournié Grospaud's performance was mediocre. When a new managing director arrived in the mid-1990s, he invited me to support his actions to redefine the company's offer. Some substantial qualitative improvements were achieved, but we uncovered the real turnaround reason in his relentless attempt to eradicate dead proposals. The rationale was simple: he attacked the justification of fixed costs. Has the sales rep or the project manager in each specific case gathered enough evidence of the potential success of the future project to commit to engage the company's resources?

Over ten years later I was discussing the relevance of this dead proposal ratio or corporate pulse with the CEO of Vinci, Xavier Huillard. The concept was still strictly followed in the Energy Division, but I discovered a supplementary reason for its being crucial to pay attention to the correct enforcement of the corporate pulse in the specific case of a listed company:

> The information requirements for listed companies represent a constantly increasing consumption of resources. Consequently, if the performance measurement keys, at the elementary level of strategic execution, are not deeply rooted in the company culture by those who are accountable for performance, the organization runs the risk of developing information systems based on ad hoc analysis, which are technically correct but not relevant enough management wise. Unwittingly, the company and especially these who generate this operational performance could defocus from the true mechanism of performance building and substitute it with some performance measurement derivatives more in line with the flavour of the month wanted by analysts. And we can note that for some companies this is becoming the corporate gospel which puts them at risk.

At this stage in SN's project, one can observe that the company had first made an appropriate use of its asset base in line with its ambitions. The strategic ambition 'Connect Brussels' had been segmented into homogenous challenges in relation to the competitive context. The company was assessing its performance at the elementary level of strategy execution with a specific focus on the most sensitive part of the business model, thanks to a new and unique corporate pulse, RoDoC, which was becoming a corporate reference for the company. SN had entered into a new relationship with its performance measurement. The next step was to boost performance.

BEAR IN MIND

Sharing information is motivating – more precisely, sharing meaningful information is motivating. The specific context of a turnaround imposes the need to reengineer information sharing in two directions. An organization must simultaneously develop a sense of care and creativity to answer the challenges of first supporting performance improvement, and secondly consolidating motivation. The concept of a corporate pulse® – a quick, accurate, management-oriented measurement of performance and its progress, named RoDoC (return on direct operating costs), and developed for SN Brussels Airlines – was the foundation of the answer to this ambitious goal. A corporate pulse is not a window-dressing exercise, it finds its rationale in the depth of a company's business model and context.

The success of a turnaround is influenced by the quality of the existing assets and their appropriateness to the desired goal. This is a prerequisite, which demands a real assessment of performance, not just sticking to appearances. To prepare for effective use of a corporate pulse, the elementary level of strategy execution has to be formed into strategic groups, defined by their similarity in the competitive context. This leads to a more accurate understanding of what level of profit can be achieved in each specific situation.

A corporate pulse is a substitution for hackneyed performance derivatives. It operates at the appropriate level of monitoring of strategic execution. In SN's case, the priority of the seat perspective was abandoned to focus on the flight. The rationale behind this shift was to prioritize for the whole management the observation of performance from the most costly part of the business model, the direct operating costs (DoC) of a flight.

A corporate pulse is a management tool for which one must find the best ad hoc expression that fits the management challenge of the organization. That is why it can be expressed either in terms of physical elements or in the form of a ratio. This result must be related to a well-known cascade of causes and consequences,

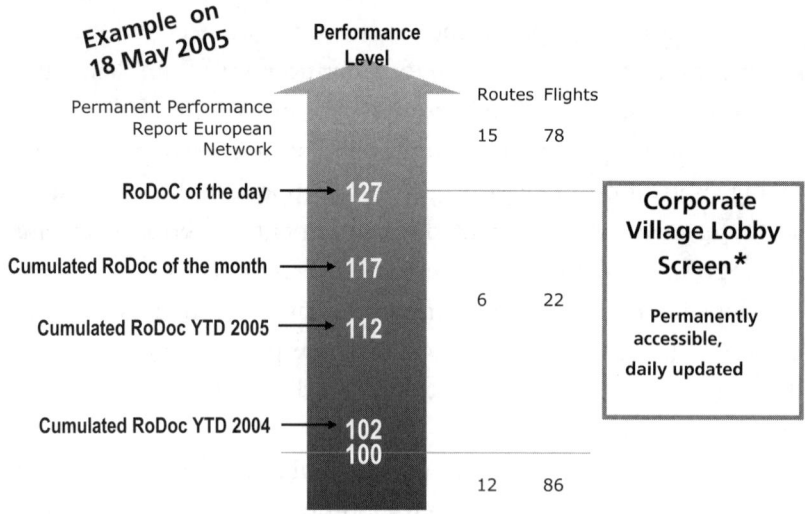

Figure 10.5 Transparency on corporate performance.

which tells the operators how they are performing and what they can correct immediately. The tool also invites the senior management to become invisible observers of what happens in real time in their operations. To do so the corporate pulse answers a simple principle of management: *approximately right now, instead of accurately correct too late.* In SN's case, this is translated into the availability each working day at 11 a.m. of an operating performance report for the previous day from the elementary level of strategic execution, the flight, to its most relevant consolidated form, the network. A daily report of the previous day's performance is imposed by the operating conditions of the sector, which deals with perishable goods. The circulation of information is transparently achieved both through a detailed report to the CEO and his direct reports and a summary of performance that is daily updated on the plasma screen in the lobby, with a well-understood colour code to interpret that performance.

PRACTICAL REFLEXES
TO DEVELOP A
CORPORATE PULSE

- Reassess the strategic relevance of the company's asset base to its ambitions.
- Create a strategic group to define challenges and their monitoring by group to optimize the use of corporate resources.
- Check your current performance measurement driver(s) and try to assess them in terms of management relevance. Are they delivering an accurate assessment, how fast and allowing which speed of reaction?
- Check on the company's P&L waterfall: which priority zone naturally emerges from the graph? Can we transform this priority zone into a 'pulse'? The rule is simple: when the priority zone is revenue the corporate pulse can be expressed in a physical measurement; when the critical issue is cost a return ratio is more suitable.
- Create an ad hoc scale of performance assessment for the company.
- Articulate with care the context for communication and the necessary debriefing sessions.

PART V

SPEAKING THE NEW LANGUAGE FLUENTLY

When an organization has developed the tools described in the previous parts, unfortunately it runs the risk of these tools being taken for granted once they have been presented and explained. Effort must be invested to make such tools as self-enforceable as possible, but that only represents a small part of the real challenge. So that the company enters into a true transformation and begins to enjoy a different style and improved effectiveness in performance management, these tools must reach the point where they become a form of reflex.

Developing a reflex with management tools is a topic that I really do like, because I have a trick in this respect. My next goal in the transformation process was to kick start a different growth dynamic. It encompasse learning how to 'speak fluently' the corporate pulse, RoDoC. I embraced this ambitious goal through the development of a very broad and deep dialogue at the elementary level of monitoring of strategy execution, in SN's case the route.

We named this process the Route Business Strategy meeting or RBS, a name that took a very important place in the new company lexicon.

The objective of this part is to show the sequence of steps in developing fluent RoDoC speakers company wide. This management development exercise represented the synthesis step of the transformation process. It was therefore interesting to observe what was tangibly visible in terms of results in summer 2005.

CHAPTER 11

LEARNING HOW TO SPEAK RODOC

FROM PROJECT TO SKILLS

This last step in SN's transformation process was a crucial one. Individual tools and their implementation were already behind us, but the challenge was now from a higher league: to nurture their synergistic implementation. If this step is missed the company transformation is not complete. It is not the juxtaposition of new tools that changes habits and behaviours, but implementing them as part of the system. That is what we will see in the following three steps, which review some components of the organization: first, how the project modifies corporate *skills*; then based on this result how these *skills* affect the *structure*; and finally how the combination of new *skills* deployed in a different structure rejuvenates the company's *style* and culture (*shared values*). At the end of this sequence, SN was a far sharper player against the competition than in Fall 2003 when the RBS or Route Business Strategy meeting was launched.

Developing a new performance information *system* to support a crucial evolution in any organization runs a classic risk, termed the 'project trap'. However brilliant, unique or disruptive a project output is, it is threatened by a quick return to the old routine because complacency is inherent to any organization.

In late summer 2003 I was just completing an initial strategy performance measurement workshop at the route level with a large team of nearly 30 people. The goal was twofold: first, make this team more familiar with the performance measurement tools and some conclusions of the first six months of the YEP pilot project regarding some specific routes; secondly, enter into a forecasting mode, by being able to assess the end-of-year performance and the forecast for 2004. Among the lessons of the pilot phase, one broke a formerly well-entrenched paradigm.

Business story 11.1: An aircraft must fly to cover its costs

After the first month of YEP observation on the Lyon route in competition with Air France, at Marseille and Toulouse two exclusive routes were reporting performances that were not so encouraging. The RoDoC was below the break-even point, ranging from 91 to 79. But our analysis did not remain at the route strategic level and made the RoDoC dive down to the elementary operational level, the flight per day of the week. Given that this took place in a pilot phase, almost any incredible idea could be suggested.

There was always a constant in the performance profile: the second flight leaving Brussels after 10 a.m. and the second in-bound flight leaving the destination at a similar time were a performance nightmare. So, playing with BEP, I suggested cancelling these two flights. In practice this meant that the first outbound flight was parked at its destination for four or five hours before returning to base. In short-haul traffic this did not seem to make sense. Nevertheless, we baptized this fleet management practice the 'long day stop', by analogy to the 'night stop' when the aircraft 'sleeps' at the destination airport to offer an early connection to the original airport.

This suggestion immediately generated a lot of objections. The most obvious one was that an aircraft is designed to fly, not to stay idle on the tarmac. That is true! When I shared this experience with Patrick Alexandre, Directeur Général Commerce International of Air France, I saw some scepticism in his eyes. I felt obliged to send him an e-mail to explain why I was making such a choice, which didn't fit the airline paradigm. This insignificant topic of a 'day stop' addresses an important management lesson for a turnaround context. The classic business paradigms can be misleading because we have an inaccurate reading of the context. Consequently we don't apply a principle of paramount importance, which can be summarized as: *What is the problem*? In SN's case, this was to break even. Unfortunately, too often we think first in terms of solutions, we don't stay open enough to what the real problem is.

In this respect I always have in mind an observation from the CEO of Intuit, Bell Campbell, to his sales people: 'If you are going to tell the engineer what features you want, I am going to throw you out on the street. You are going to tell them what problem the customer has. And then the engineers are going to provide you with a way better solution than you'll ever get by telling them to put some dopey features in there.' This realistic attitude is a good one, but unfortunately it is not widespread among companies. The easiest way is always to think more in terms of solutions than problems.

Regarding SN's long day stop, I was suggesting a diversion from the sacred cow principle that aircraft need to fly. They need to fly when the cost of the flight matches the contribution of the legitimate accessible market. So when that is not the case, one has to play with the four ways to deal with BEP: increase revenue, cut fixed costs, increase productivity through variable cost optimization, or finally change the mix. What was available in SN's specific context? Neither of the last two proposals, it was too early. Here, we needed to act urgently. Dreamers would suggest increasing revenue, but this rarely happens unless you have a secret to offer to the market. So what you can do has to be costs related. The 'long day stop' was just an ad hoc answer, nothing more. But a turnaround is made up of a lot of ad hoc answers until the business model is revalidated.

Once tested, this decision demonstrated that it was full of business common sense and the traditional objections imagined by structural detractors didn't last very long. The main objection was that passengers would walk away to another airline. SN was exclusive on Marseille and Toulouse. On Lyon with the day stop it matched the flight frequency of its competitor. So SN kept its passengers and diminished its costs. The next YEP report, in April 2003 after only 14 days of implementation, reported that Marseille was at the BEP, Lyon was even exceeding it, and Toulouse had gained 10 points of RoDoC, nearly closing half the BEP gap. Three months later all the routes were above BEP and two of them had even entered the profit zone.

When I proposed organizing a large workshop involving all those concerned with route performance, the context was quite favourable, because the internal ambassadors had started their job of promotion. Once somebody has obtained a profitable route in their portfolio, they are not ready to see it falling back into the red again.

The long day stop delivered an important additional lesson. The correct choice of flight frequency, influencing the total direct operating costs on a route, is critical, because sales development cannot compensate for extra DoC due to an inappropriate level of asset deployment. So fixed costs are crucial – an obvious point, but one that is too often forgotten. This rule brought us back to the appropriate choice for asset use, which influences the profitability of the operation. So the long day stop immediately killed the culture of marginal costs with inaccurate figures. When the average European network DoC are around €5200 with 63 per cent variable DoC and an average passenger contribution around €115, the minimum average number of passengers per flight in order to cover the variable DoC is 28. Consequently, it is above 28 passengers that the concept of marginal revenue begins to become relevant. It is not at the first additional passenger, as some salespeople commonly said. This business reality has not been appreciated and it operated like a cold shower on many preconceived ideas. A very talented, classic airline-style salesman took many months to get into this correct cost accounting way. And many times he rang me to check whether his proposal fitted with 'your YEP things', as he put it.

Another conclusion is important to address. From an asset management point of view, the aircraft that was no longer needed on the route was grounded. The fixed part of DoC was corporate costs, but it was decided not to have it supported by the route. The goal was to reach break-even through the implementation of operational decisions. A substantial part of this result was achieved thanks to asset use rationalization. What was important was that the people responsible for this decision read its direct impact.

Secondly, it was also wise to see the costs of unused assets increasing. This forced the senior management of the company to question whether the volume of assets was appropriate for its strategic mission, 'Connect Brussels'. We saw earlier that SN was fortunately born with the correct asset base in terms of quality.

This deprived it of the need constantly to consider whether this asset base was also appropriate in terms of volume. And in the case of an excess of assets this demanded a rationalizing decision, which concerned all the constituents of the fixed costs part of the DoC: aircraft, crew and maintenance. If it was not implemented with enough courage, people were just paying lip service to this form of management discipline. I fully admit that it is not always easy, but it is the heart of the CEO's mission to anticipate this kind of situation when the business is favourable.

> **YEP pulls the walls of the different clusters down.**

So the YEP report was starting its role of pulling down the walls between the different clusters, mainly sales and production. In this process of bringing together resources and talent, I also received an unexpected push, which spoke to people's pride. The question of fleet renewal was on the agenda of the company virtually from the relaunch. One of the executive committee members, Peter, was assigned to this topic and was preparing alternative cases for the committee. To do so he was visiting aircraft manufacturers and one day he told me that he was planning to organize a meeting with the founder of Crossair, Moritz Suter, who had been CEO of Swissair for just three weeks and left when he saw the company's desperate situation and that nobody was ready to change. Moritz Suter was also known to be very insightful in terms of fleet renewal. Crossair was a model as an aircraft first user. When Peter invited me to join him for this visit I gladly accepted.

We spent all day in Basel. Moritz Suter gave us a private masterclass on fleet renewal. I was very happy with what I had learned and during the discussion I showed him the YEP report and explained the rationale behind the approach. A few days after the meeting I received a thank-you note for *Organic Growth,* which I had given to him, plus a very positive comment on YEP:

'It is the first time I have seen such a document and it is a very interesting, refreshing approach to performance assessment.'

So the concept of YEP, supported by internal success such as on Lyon and Marseille and by welcome outside help, was encountering less and less resistance. The workshop in August 2003, including all the destination managers, the Belgian sales team and the network, also went well. We played around for two days with simulation tools in workshops arranged by groups of destinations. Once the workshop was over, the team was very excited because it was approaching performance from another angle. But simultaneously, I felt very concerned because for the first time I had a full picture of the difficulties of the next step. My challenge was the following: We were not yet in the heart of the transformation process. Some tools were available, but none of them had reached autonomy in use. So the next challenge was to create the context that would help develop autonomy in use or transform the project into individual skills that could be spread company wide.

In late summer 2003, the European network workshop was a fair success, but this was not enough as long as a synthesis of the discussions per route was not available on the desk of each of those responsible for route performance. They needed to receive a milestone in their new skills development, which could be used as a reference point. The rule was a simple lesson in humility. The synthesis had to be done and there was only one rule to ensure homogeneity in its structure: write a synthesis of everything. And I did it! If at this early stage of the change process you plan to rely on the collaboration of the team, you are making a mistake. My philosophy in change management is not a jump into the unknown. The map must be drawn beforehand and this synthesis was a draft of a consistent map with the same format and the same theme, crafting each route's winning formula. My ambition was to leave the sterile discussion of past performance and in September 2003 to make the staff dare to estimate the results at the end of the year in the same presentation format.

The organizational sleight of hand at this stage of a change process is to transform the project output into corporate staff's skills. A Chinese proverb clearly gives the tempo of the challenge: 'When a man is starving, you can give him a fish or teach him how to fish.' SN's new performance management focus with all its bells and whistles was like the fish, and the ambitions of a real organizational breakthrough would be reached only when SN had been taught how to fish. This would then create managerial autonomy, which would transform a group of senior executives into active strategists.[36] Another analogy can be offered for quality management fans: the challenge is the same as how to find the wedge that avoids Deming's progress wheel moving backwards. So the next step of the challenge could be summarized as making people speak RodoC fluently.

FROM SKILLS TO STRUCTURE

Former experiences of cross-fertilization provided me with some very important lessons to stick to in order to transform a project output into skills. One of them stemmed from the automotive industry. Why do some automotive manufacturers achieve superior performance in terms of faultless production results while almost the whole sector is using the same Japanese-influenced performance improvement tools? A clue is that only a few of them have developed outstanding skills in running transverse projects, while others just enjoy the illusion of doing well or remain structurally outpaced.

---

[36] By strategist, I refer to the definition given by General Von Molkte, the organizer of the Prussian army: 'UNDERSTAND the meaning of events without being influenced by popular opinion, attitudinal changes or one's own prejudices, TAKE quick decisions, TAKE action without being intimidated by a potential danger.'

But this capacity to work together, whichever departments the participants came from, was translated into the details of the structure of those that were successful, not of the corporation itself but of the strategic business unit. This is what I was able to observe when I gave a team from SN the opportunity to visit one of Renault's most productive car-assembly factories, at Douai in the north of France. For one day, SN's maintenance team observed and discussed with senior executives from a stamping and welding department in Renault a difficult organizational topic: how effective Renault was at achieving the targeted performance improvement goal, thanks to a department support body including production, engineering, maintenance and quality technicians, where the whole team reported to the department head. In this case, the distinctive skills of running transverse projects had been integrated and pushed down to the lowest possible relevant level to optimize the speed of reaction, due to their capacity first to deploy their individual skills and secondly to work together on a common strategic goal.

> **A transverse team at the lowest level of monitoring of strategic execution is a characteristic of a superior-performing organization.**

SN's challenge was of the same nature, creating a transverse organization, but with a slight nuance. Given that the structure could not be immediately modified, an ad hoc dialogue platform was created for all those involved, a group of more than 20 people who could influence strategic business unit (SBU) performance. These performance influencers had the opportunity to meet and discuss within a specific structure the route strategic diagnosis and its profitability improvement action plan (see Figure 11.1).

Creating this discussion platform was one thing, but making it work not once but systematically was really a matter of discipline and commitment. The first key point was to make it clearly

Figure 11.1 Stakeholders of the Route Business Strategy meeting (RBS).

understood that in this meeting the rule was 'one seat, one voice' and that there was no hierarchy. But in cultural terms this was not obvious. The ambition for the RBS meetings was huge: to develop 'deep insight into the trends, in demographic, lifestyle business structure to unveil new competitive areas and superior performance for SN'. Its success was only to a limited degree down to technique. I repeated a hundred times and ended up being listened to when I said that foresight does not come by being a better forecaster, but by being less hidebound, or in other words more open. This is made up of a good dose of mindset, a substantial dollop of systematic work and, if one is lucky, some drops of insight.

In the development of the RBS, I was lucky because one of the Network VPs who was returning from maternity leave was ready to embrace this challenge beyond her normal workload. Christine was able to motivate some of her most talented and key colleagues. So the change process started to find some additional weight when it turned into a team mode. A new wind was blowing and this is made clear in Figure 11.2.

I continually repeated and stressed, 'Analysis is good, implementation in day-to-day business life is better. The ultimate goal of RBS is *bottom-line impact*.'

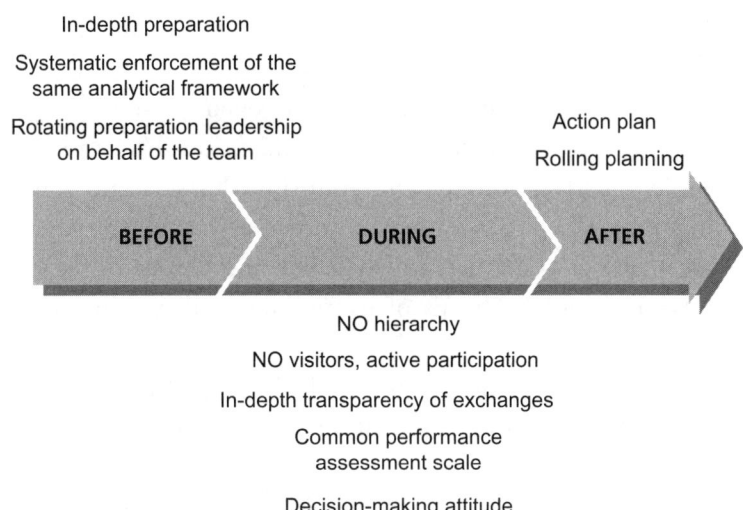

In-depth preparation

Systematic enforcement of the
same analytical framework

Rotating preparation leadership
on behalf of the team

Action plan

Rolling planning

BEFORE   DURING   AFTER

NO hierarchy

NO visitors, active participation

In-depth transparency of exchanges

Common performance
assessment scale

Decision-making attitude

Figure 11.2   Rules of the game for the Route Business Strategy meeting (RBS).

The discussion platform of the RBS meeting pulled down the walls between two key departments, sales and network. Secondly, it forced people to enter into the same systematic process of strategic diagnosis in each business unit. The investment in terms of people mobilization was huge, with more than 20 participants. But the in-depth preparation and the strict meeting discipline enforcing the correct individual attitudes quickly transformed what could have been a mess into a collective learning process for each member. The result, using the words of the participants, was a 'deep–broad–hard-facts based cool-mind action–oriented strategic diagnosis followed by a strictly monitored action plan'.

Today the results are clear, but the management development process was not as obvious as one may imagine. The first RBS meeting in Milan was a kind of nightmare. We met in SN's crew hotel, which was downtown, but the conference room was too small for the more than 20 people packed into it. The discussion time per route was not kept to and we were unable to discuss some routes. When you observe this situation for the first time you don't bet a lot on the future of such a venue. But 18 months later the outstation meetings were all conducted in the airport building and the catering for them was even subsidized by the airport authority. The meeting started and finished on time, with the whole agenda covered. The company did not know where the difference lay, but as an outside observer I could unquestionably say that the RBS meeting had become a part of its genes. The RBS had left the hands of its designer and was becoming a collective corporate possession.

Over less than 18 months the tools developed in the first few months had become organizational reflexes embodied in corporate

Route Performance Ambition

	Market track record	Nature of the route. Which growth?
	SN's past season track record	SN's offer and performance. SN's relative performance against route evolution
Area of observation and analysis	Partnership's ambitions	Rationale of the agreement Are the benefits fairly shared?
	Competition's ambitions	Who is the enemy?
	SN's targeted positioning	Leader Active challenger Follower
	Strategic objective	Quantitative: RoDoC, Fleet contribution
Deliverables	Mobilizing theme	Qualitative: 'Stop low-cost growth' Eye on the ball
	Action plan	Measures/impacts

Figure 11.3 Route Business Strategy meeting (RBS) roadmap.

skills due to the RBS structure, and were turning into sustainable and systematic practices.

FROM STRUCTURE TO STYLE AND SHARED VALUES

RBS was deployed thanks to an ad hoc structure that provided a dialogue platform for development of the skills. These skills became a element of SN's culture due a very interesting combination of commitment to preparation, a cool head or lack of emotion when dealing with results, confidence in the reliability of data and its structure of presentation, and transforming performance measurement into fun for the whole group.

An effective driving range

Over the past 18 months each SBU, a route, had been through this performance-filtering process three times. It avoided becoming routine because of the quality of the in-depth preparation techniques, which operated on a rotating basis, each time involving a different department as team leader of the RBS module. This was the critical point of this transformation mechanism. At the beginning during their preparation people were coming along with a lot of bits and pieces. The first step was to structure the outline of the discussion, but very quickly the team saw that it was not enough. What was critical was to be perfectly prepared according to a systematic, analytical process.

Again, I observed that if one wanted to get a different result from these committed people, it demanded more than just setting up an agenda and giving some instructions for preparation. One needed to commit oneself in the heart of the effort with the preparation team. Consequently, I proposed a follow-up preparation schedule and spent long hours supporting each sales rep in the discovery of their new role. But once the analysis had been done, I made them jump the classic hurdle of diagnosis formulation. At this stage, I knew that the war was won because this synthesis could become the reference point thanks to the sales rep's ease at playing with the data. This developed the confidence of the group and prevented inconsistency occurring in the discussion, because it gave a tempo to their professionalism.

> **Commit yourself to being in the heart of this transformation process, you cannot delegate.**

The trick behind this result was simple: a sense of detail. The preparation ended with a systematic rehearsal. This was a rehearsal for everybody including me, so I attended all the rehearsals of all the presentations of the Spring 2005 session. I did that not for the conclusion of the analysis nor the synthesis of the diagnosis, but to help these people operate at their peak. This was done with people ranging from below 30 to more than 60 years old and they all performed well. This was so clear that the Commercial Director Belgium did not miss the opportunity to tell them how proud he was of their new attitude in an e-mail in May 2005:

Dear all,

During the last Commercial Meeting, Philip expressed his gratitude and satisfaction regarding the quality of work all of you have delivered during the first 2005 RBS sessions.

I wish herewith to congratulate you all for the dedication, efforts, extra workload you all performed. Your professional know-how has now reached another milestone and will bring additional added value to the company. I was very pleased to hear from some of you that after this exercise your minds shifted from a wait-and-see position to an eager-for-more position.

Thank you again for your trust, confidence and involvement.

Best regards,

Etienne

The preparation lit the spark for the team and its performance. This helped transform those who were sceptical into committed supporters and their skills into a form of corporate culture. This was a genuine moment of truth for me about how the cement of these ideas was solidifying.

I consider that I was very successful at increasing the emotional intelligence of these executives. This was not a form of manipulation, but a form of respect for what they could really do once the SABENA routine had been broken. After more than 20 years of management development, I remain a strong supporter of getting outstanding performance from 'ordinary' people. I believe in their capacity to achieve spectacular transformation by instilling the right stimuli. After SN I took the challenge of committing myself to the turnaround of Caribbean airline BWIA for this unique reason. I wanted to tackle the limits of my own belief in how to achieve superior performance with 'ordinary' people who have been anaesthetized by years of never-ending routine. In doing so, I consider that I always use four leverage techniques:

- Creating meaning and purpose to develop full confidence in one's own talent.
- Improving skills to influence one's mindset to turn on one's emotional intelligence.
- Putting in place structure, recognition and reward.
- Selecting leaders to serve as role models and cascade the process.

Individually these four transformation gears are ineffective, while they are a highly efficient breakthrough when intertwined and monitored accordingly.

This mechanism very quickly multiplies the number of staff members who think and speak RoDoC and naturally play with route performance leverage points without any apprehension.

No more emotion or passion but facts

SN reached the point where the project output became skills when it was noted that this group of senior executives assessed any performance result without any emotion but with the commitment of empowered-accountable managers. This was a very encouraging, substantial intermediate result in the evolution of company performance management. In the management conference in spring 2004, which took place at the end of the first RBS session, I remember Rudy, the Sales VP Europe, saying publicly that the RBS process had a big impact on his troops:

> There is no more emotion about performance information, the results are what they are. The sales reps and the country managers are able to explain where this performance is coming from, and paradoxically in this explanation delivered by sales-minded people each of us now starts addressing the results from the costs side, saying here is the amount of contribution. I jointly with my colleague at the other end of the route have to cover the DoC corresponding to a market-relevant proposal of our best assessment of the optimal number of frequencies.

In his comment, Rudy was also sending clear signals that the RBS meeting was cementing the company structure. Country managers now were observing performance for which they were jointly accountable. Therefore the company with this group of managers was operating one step beyond its traditional perspective, sales targets, and was enjoying joint accountability for profitability. The concept of an elementary level of strategy execution had become actionable throughout the organization, because one could identify a team associated with each elementary unit of strategy execution. This observation was critical, because real transformations are achieved when one can see a structure associated with the way one expects to see matters evolving. At SN there was no organizational chart of the elementary units of strategy execution, but these teams were effective because the staff involved in the challenge knew clearly which team they belonged to, who were their teammates, and which behaviours and attitude were expected from them.

In this effort to build a team at the elementary level of strategy execution, the role of the functional staff was also critical. It was clear that on the route there were some functional people who organized production, called in this case the planning managers. To people in the field, they were perceived as a form of necessary evil with whom to fight constantly, because they were seen as people who always said 'no' from their ivory tower, disconnected from business realities. The planning managers in their turn knew the song by heart, and in their view their optimization rules were looking at the business from a broader perspective. I saw tons of forwarded e-mails showing clearly that nobody was really reading what was written.

RBS put an end to this form of sterile dialogue where the answer was the result of who had perceived power, because it was the first time that their relationship was oriented towards a common objective where there were no losers or winners, only one objective: corporate performance. There is a big difference between a common view and a juxtaposition of views. From now on the result would go in the direction of what was gauged as optimal for the company from a corporate perspective, not from the most influential department's one. The deciding factor was the elementary level of what was in the interests of strategy execution. This found a practical translation in the rotating preparation process, which was driven by fair rules: over 18 months each of the three sides – destination, origin and network – led the team in preparing for the RBS meeting. This exercise forced them to think in terms of the corporate interest. The blinkers were removed.

One route, one binder

The team was accountable for preparing for the RBS meeting. But supporting the change process required a different way of looking at information. The company's information system, which was largely influenced by SABENA's way of dealing with data, was not ready to deliver it in a format that fitted the information requirements at the elementary level of strategy execution. To structure a correct path through the company's information jungle, we developed with a team of my MBA students an ad hoc tool: the route binder. This 'pre-chewed' the data in order that managers could concentrate on the meaning of the analysis and didn't have the excuse that the information was not available.

In this binder, which was structured in sections, one could find everything in terms of markets, offer (schedule), costs structure, revenue structure and BEP, a journal of route events (sales, competition, price) and YEP reports (weekday, weekend, consolidated, by point of sale). There was one binder per route and each month Silvia spent a lot of time keeping the portfolio of binders updated. In this task she was helped by the performance measurement team and its database whiz, who had the unique talent of creating tons of reports with an incredible level of flexibility, which made you believe that any report was a customized one. His sense of anticipation made our task lighter. Nevertheless, updating all the binders and e-mailing all the stakeholders represented a significant amount of energy.

Business story 11.2: 'What about a route binder quiz to warm up today?'

In a change process a tool is nothing unless it is used. Everybody was pleased to receive this new set of information, but they didn't know that I hate waste and I always pay a lot of attention to making sure that one is not threatened by this poison. The first RBS meeting was a pilot project, with a trial-and-error flavour divided into two sessions. The first goal was to begin this new form of discussion. The 'strategic' discussion at the route level was not so obvious and to be sure that a certain amount of these fundamentals had been correctly digested, a second session was planned six to eight weeks later. Meanwhile everybody had received one or two updated issues of their respective route binder.

I welcomed the first team with something that surprised the whole group. Some of them called me professor because they knew that I was a senior lecturer in a French business school and that a team of my MBA students were there at the beginning of the project. But they had forgotten that a professor controls knowledge. That was why I prepared a short quiz with 12 multiple-choice questions about the information content of their respective binders dealing with the concept of RoDoC. I asked, for instance, what the DoC of their respective route was. I was interested to observe their excuses for not answering the question correctly.

One of the German sales reps, with his country's natural reserve, swallowed many of his wrong answers. But it was too much, he could no longer support what he perceived as a form of humiliation. He assessed the rules of the game as unfair. I was surprised that he did so, because each month all the route joint managers were sent the updated data for each route binder. And our German colleague was one of the addressees. His question was: 'I understand that this information is important, but where can I find it? I have just a vague idea of what the response may be and the multiple-choice question asks for something accurate.'

Surprised, I replied: 'In your route binder.'

'Do I have a route binder?'

'But you are supposed to get it each month . . .'

Then a third voice entered the dialogue, saying: 'I keep this document because I want them to concentrate on sales.'

It would have been ridiculous for me to be furious. The country manager was just telling us that as God is in the details, management change also lies in the scrupulous monitoring of obvious things. Did it mean that by keeping these pieces of information he was demonstrating a form of information possession power? Not at all, this country manager was quite new and had not found the time to digest all the aspects of the evolution demanded and he did not want to expose himself to management situations that he had not yet assimilated. Consequently, it sent a message to the RBS managing team that the context had not been assessed carefully enough for the change process not to run the risk of unwittingly being stopped for ridiculous reasons. It is never a waste to monitor obvious things such as how these documents are shared with subordinates. The cost of missing such a step is almost unaffordable: it undermines the confidence-building process.

The objective of the binder quiz was that the whole team of performance stakeholders should master, like a reflex, a set of critical and updated information in order that any suggestion could be addressed in terms of impact on the BEP. Some people whom I caught on the wrong foot told me that they knew where this information was. But this is not appropriate if the objective is to create a new tempo for a company. They are managers, not librarians. So often I have heard people speaking about a business model, but they don't see its consequences. A business model is like a foreign language, you need to speak it fluently if you have the objective of operating in a different cultural environment – you don't want to be like a fish out of water. Everybody agrees with this image, but when one addresses management topics one is not spontaneously observing the same eagerness for ease or fluidity.

This trick was just an additional practical translation of what I described in *Organic Growth*, when I explained that my former MBA classmate, Thierry de Kalbermatten (who in the early 2000s was MD of the world-leading Swiss packaging equipment manufacturer BOBST), got rid of functional support to have sale reps who in front of the customer were able to answer at least 80 per cent of the customer's questions, ranging from equipment performance and waste management to leasing. If this is true for salespeople, it is also true for managers, who should immediately on the back of an envelope be able to gauge whether performance is good or not.

Performance measurement can even become a theme for jokes

One important question in a change process is how to prevent rejection of the change. Above, I explained that the process is highly time consuming, with support and rehearsal for instance. But there is another important question: when do you know that the knowledge and the new business reflexes are in place? The answer always catches you on the wrong foot because it comes without warning.

This happened to me in late March 2005 at the RBS meeting in Geneva for central Europe. The RBS followed an incentive weekend for all the European country managers in a Swissski

1. Thou shalt not have a RoDoC of less than 63

2. Thou shalt not remove BHX from the top 3 most profitable routes

3. Thou shalt not ignore the shit list

4. Thou shalt only look at the cutbacks on a Saturday

5. Thou shalt always cover variable costs

6. Thou shalt not covet thy neighbour's RoDoC (especially GVA)

7. Thou shalt worship only at the church of YEP

8. Thou shalt prepare thy RBS thru the sanctity of the Professor

9. Thou shalt not confuse YEP, BEP, YAP, TS, PS VARDOC FIXEDOC, RoDoC

10. Thou shalt believe in the virtue of the holy trinity – analysis x 3

Figure 11.4 SN's 10 commandments.

resort, Los Diabkrets. When the team of European country managers, suntanned by two days at 3000 metres, arrived at Geneva airport for the RBS meeting, they told me very excitedly, 'Yesterday evening we prepared a specific slide for this RBS session.' It was on SN's ten commandments. 'You'll see, professor,' said the UK manager, laughing. This slide, which has become famous throughout the company, is reproduced in Figure 11.4.

This was a jewel, because it expressed what they had in their heart. But it was also great because the vast majority of what we had been insisting on for many months was in the document. And this was done on the back of a menu saying the Victoria Hotel, Raclette – it is easy to imagine the atmosphere of fun, helped by the Fendant, the Swiss white wine values from which they wrote something so well tuned. This was an encouraging signal that we were on the correct track. I mentioned on page 200 the spontaneous comment on the YEP report from Moritz Suter, where an outsider with in-depth knowledge of this sector was sending an encouraging comment. This time it was coming from inside. This had to be read as a signal that the transformation process in terms of performance measurement was culturally rooted in the company. SN had rebuilt a stealthy competitive advantage, because in the RBS sessions the debates were real and generated positive friction for the good of the company.

The whole performance measurement process from diagnosis to analysis was under control and even the style sent positive signals: RoDoC had become a substantial brick of the culture of this just-reborn company. But if RoDoC helped the whole company assess with lucidity what had been done, RoDoC and the RBS had to take up the more ambitious challenge of *anticipating* performance. In terms of culture SN now knew how to live with this recommendation from Gary Hamel:

> Building industry foresight demands that senior management be willing to move far beyond issues on which it can claim a expert status. They participate in debates about the future as equals, not as omnipotent judge.[37]

Before addressing the last step of this corporate management development process, it is worth seeing what was happening to performance in this company with all the tools implemented, when RBS had become a best demonstrated corporate practice.

[37] Gary Hamel (1994) 'Seeing the future first', *Fortune*, 5 September.

CHAPTER 12

IMPROVING THE
WINNING FORMULA

LESSONS OF THE PERFORMANCE
MANAGEMENT TRANSFORMATION

At this stage it is important to show from different angles the quality of the results and how they were an expression of genuine management health, which allowed SN to claim full membership in its sector having left the ashes of the SABENA bankruptcy only 40 months previously.

· At the end of summer 2003, before the launch of the RBS sessions, we ran a workshop, the objective of which was to identify the winning formula of each route. After 18 months of in-depth discussion of performance, it is fair to conclude that the actionable winning formula had been sharpened. I propose to outline what SN could count on at this stage of its corporate history. Thanks to the in-depth exchange in this new virtual structure, some old airline-sector paradigms had begun to be systematically questioned and quite naturally replaced by a fresh way of handling performance improvement. The main break-through concerned the systematic enforcement of prioritized and coordinated performance improvement actions.

Given that the airline industry is a capital-intensive sector combined with a high level of variable costs in its operations, ranging from 64 to 69 per cent respectively in Europe and Africa, the importance of accurately adjusting the asset deployment level – the correct number of frequencies to operate – to SN's legiti-mate accessible market was paramount. In the overcapacity context of SN's operations, this skill was even more critical. By disregard-ing it, the organization ran the risk of never aiming at a substantial enough fleet contribution, always making costly trade-offs in favor of ill-founded customer expectations. The argument that this cutback was not possible for customer service reasons was no longer heard after two sessions of RBS meetings. The cost pro-vided the tempo, because the company was discovering in the highly competitive environment of its European operations what its legitimate accessible market was.

> **Costs influence the tempo – frequencies are crucial.**

In a turnaround situation, overfocusing on the value side is not a very relevant choice. Having been confronted in the devel-opment of Vinci with many cases of turnaround (see page 18), Xavier Huillard's management philosophy on this question is unambiguous – 'Downsize, we cannot bet on an exit by the top,' meaning that one can't depend on a revenue increase. This obser-vation was shared by Rob Kuijpers, who added a nuance: 'Maybe in high tech it is something possible.' This is what Jean-Marie Descarpentries successfully tried at Bull in the mid-1990s or Steve Jobs when he came back to Apple. But one needs to have luck to find jewels in the company's CAD-CAM system as good as the iMac.

This awareness of the BEP created the conditions for optimization, where efforts to develop and manage the availability of a sharply competitive price structure would not spoil the restructuring effort. SN's most aggressively priced fares on its European routes were one of the most popular measures of securing volume in the cabin and of developing a 'best buy, best value' perception, while securing higher-contribution passengers at the same time thanks to an effective revenue management policy. But this had never been a reason to deviate from the sense of customer service with which the company had been relaunched. That was one of the three conclusions of the senior management customer survey in late fall 2002. SN Brussels Airlines' sense of customer care was regularly reported as one of the major breakthroughs from the previous epoch. Incidentally, this did not mean anything sophisticated: what was different from SABENA were the smiling faces of the cabin crew. SN's travel experience value has been the object of constant care, both in terms of active commitment by the cockpit and the cabin crew and in terms of rejuvenating the service content, such as keeping catering free in economy class. Peter Davies had always considered cutting this kind of service as a 'penny wise, pound foolish' attitude in SN's specific context.

The coordinated efforts influenced by an insightful production plan were turned into a winning formula combining the following:

- Fine-tune the right capacity.
- Play with elasticity of passenger contribution growth.
- Manage fuel surcharges at the right time.

The effectiveness of this winning formula was observable in the year-on-year deviation analysis in Figure 12.1, which stresses how this formula was opportunistically implemented to meet the increasing difficulties of the operating conditions, in particular the increase of DoC due almost exclusively to fuel price rises.

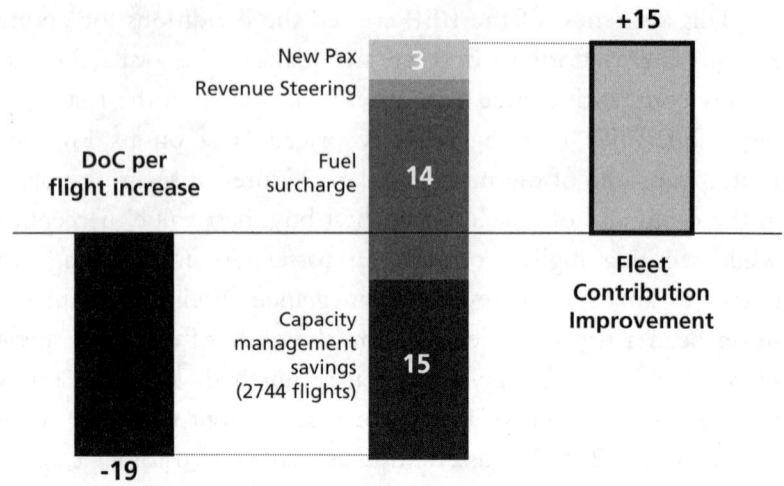

Source: YEP Report

Figure 12.1 Year on year (Q1+Q2/04–05) operational performance improvement (€m) on the European network.

Moreover, it brought enough dynamic to keep boosting performance in this general margin–chopping context. The performance-building mechanism had become simple.

SN had two route profitability improvement gears:

1. Direct operating costs, influenced by
   - evolution of unit DoC
   - total flight frequencies
2. Total passenger contribution, influenced by
   - evolution of the number of passengers
   - evolution of the average unit passenger contribution (revenue steering)

Performance improvement based on the above winning formula showed encouraging results at the basic strategic level of execution, the flight. Figure 12.2 shows that over the last six months the gap from operating margin BEP (the result after overheads

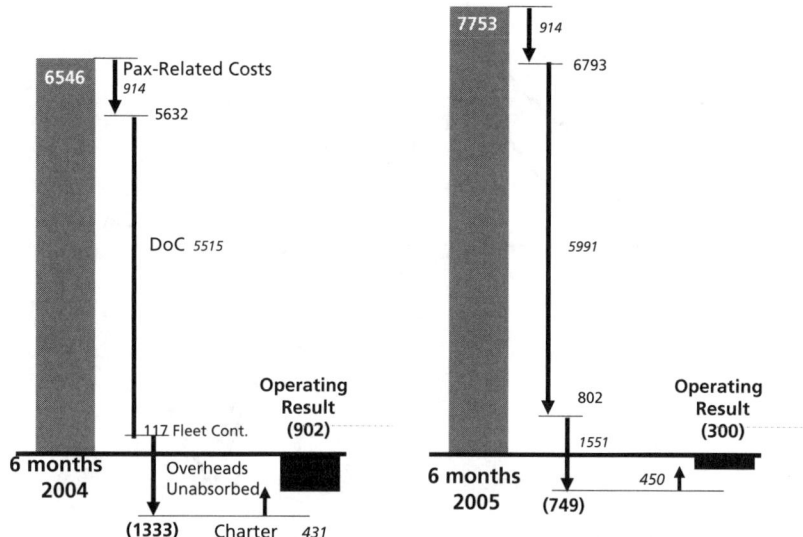

Figure 12.2   Year on year (Q1+Q2/04–05) operational performance per flight, European network.

absorption) had been cut in three. Other revenue remaining stable, this demonstrates that it was in the heart of its business where SN found its own dynamic for performance improvement, leading to a sevenfold increase in fleet contribution per flight. Finally, one can also note that the unabsorbed fixed costs remaining stable meant that the company had identified its structural level of asset redundancy.

The evolution from seat to flight was not an end in itself, but the first step of a management attitude to prevent complacency by remaining in a continual search for untapped sources of potential route-profitability improvement. The rationale behind this approach was a simple management principle: 'When a performance sends signals of a ceiling effect, break it, break it, break it, which means resegment it.' By analogy in the case of route performance improvement, the way was to change the angle of observation. A day-of-the-week perspective was added to the

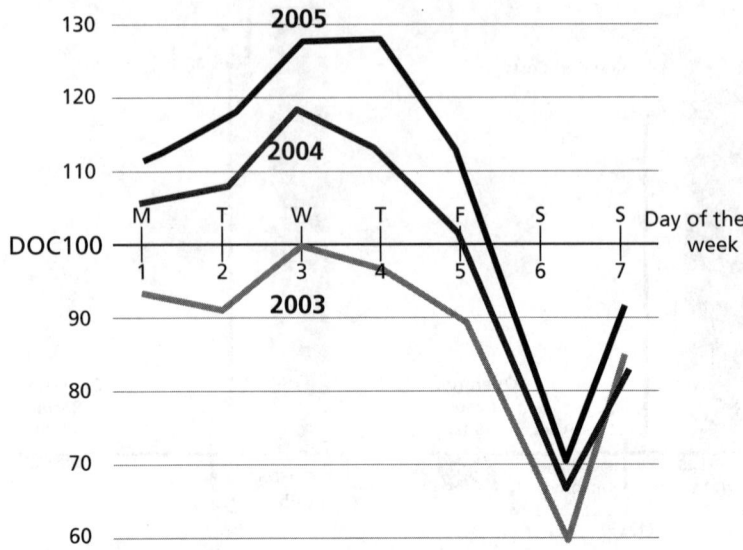

Figure 12.3 Comparison of RoDoC performance per day of the week over the first five months.

analysis. This sent the same signal of quality of performance improvement as one can read in Figure 12.3. But this additional angle of observation brought an interesting nuance, which required the efforts and the creativity of the team to be focused on weekend improvement. This led to the suggestion, which has not yet been implemented, to withdraw for instance from some southbound operations when a code share partner was operating.

In summary, over the weekend losses dropped by 65 per cent, but simultaneously the weekday performance improved by 120 per cent, four times more than the weekend performance. Through this form of observation the company received additional confirmation of its market positioning: 'Connecting Brussels' in Europe meant business-related traffic. But this way of boosting performance was now in the genes of the management team, as Business story 12.1 illustrates. Behind the result there was a form of perpetual movement, which was now actionable in the company.

Business story 12.1: Dregs of Iberia!

The lesson from the performance improvement process was to fight any form of complacency. To some extent SN was a victim of its own success. Since the first session of RBS in November 2003, we all knew that the weekend was a red-hot topic. But it was interesting to observe that the analysis was for a while biased by some forms of fallacy that were justifying the continuation of the service. For instance, it was argued that we needed a heavily loss-making flight for African connecting reasons. This appeared to be only barely correct when we got the real data. Then, the charter operations from some European airports during the weekend gave the feeling that they were compensating for the poor performance of the scheduled operations. But the results of the charter operations were observable at the corporate level, not at the route performance level. This provided another level of vagueness that kept justifying a prolonging of the pain. And finally, the stage was reached where there was nothing else to do but stop flying, because the performance was falling to a RoDoC zone that was absolutely insane, such as 50.

Nevertheless, this did not occur as the result of cool-headed analysis. No, there was passion! This passion was due to the Spain country manager, who was a genuine entrepreneur and who had requested some additional cutbacks (a decision to reduce the number of flights on a day or over a certain period) during the week of 1 May 2005, because he knew from experience that the traffic would not be good. There were something like five e-mail exchanges, finishing with a ridiculous answer from somebody from revenue management: 'We are keeping things as they are because the new Belgium sales head has just arrived and it is too early for him to make this kind of decision.' So the Spain country manager swallowed his comments and let the aircraft fly to Madrid as planned on this specific weekend. He was a hard worker and he knew that revenge is a dish best eaten cold.

When the topic arrived on the table during the RBS meeting in Spain, quickly the scope of the problem exceeded the reason for the correct decision not having been taken. Peter Davies, who has a great talent for always seeing a silver lining in the worst situation, would have found this one appropriate, because from a modest tactical planning perspective the all-RBS team was thrown into a strategic issue. Five SN flights operated with SN aircraft during this weekend, carrying 253 passengers or 51 passengers per flight, which is not bad with respect to the airline's positioning during the weekend time. But the RoDoC was 45. The worst point was the average passenger contribution of €72, which was €15 below the direct operating cost per seat, just to give another indication of the situation of poor performance reached. It was quite simple: the company was facing both a capacity and a positioning issue. During the weekend Iberia, the code share partner with larger aircraft (Airbus 321, 200 seats), its commercial tools (direct and indirect sales) and its positioning, could fight to develop a correct product with an appropriate volume to achieve a decent job. But this was not SN's position with its 100-seaters. Intrinsically, the southbound low-cost positioning was not a string to SN's bow and its traffic was just the dregs of Iberia's. This could not be maintained for much longer.

At this stage, there was no way to escape the accuracy of the diagnosis of SN's performance filters. But even in a very positive dynamic of performance improvement, one must force oneself to wonder whether one is not missing an opportunity. Again, I feel very comfortable with Andy Grove's famous phrase 'only the paranoid survive', especially at this stage of the company's life. But this not just a good association of words, it is a form of repellent against complacency, which is just here to remind us of the complexity of any business situation.

We had accumulated a certain number of lessons regarding performance measurement techniques, and in each case one could note the critical role of the whole management structure. Therefore we need to find a way to spark emotional intelligence among these managers and break their previously well-entrenched business paradigms. Finally, the difference in terms of cumulative results in the first six months of 2003 and 2005 was €30 million fleet contribution and 32 points of RoDoC. But there could be a flipside to these objectively encouraging results, which would take the form of a question: What is the origin of this result, a rejuvenated dynamic or a juxtaposition of one-off measures?

This quality of results did not occur overnight. I could observe the transformation, but in my role I needed to anticipate what was behind the result. That is why in this positive context I insisted on sorting the performance improvement into one-off measures versus what was recurrent. One delightful objective was to generate a dynamic of perpetual growth. We can all anticipate a positive future, but no tree has ever reached the sky and each business model is born with its own limits. It is good management practice to recognize this in order to avoid transforming success into frustration. Continually analysing the origin of any improvement is a sound reflex of good business common sense.

MONITORING THE ORIGIN OF PERFORMANCE IMPROVEMENTS

In early April 2005, based on the YEP report, I wrote a memo to SN's CEO headed 'One swallow does not make a summer nor a quarter a year'. I could also have said, nor does a profit make a sustainable turnaround. The end of the year on the European network was almost exactly in line with our last estimate at mid-September. The difference came from the good news of a temporary release of tension on the oil price, which increased by 14 per cent instead of the 16 per cent initially estimated. To offset this extra operating cost, the flight programme was cut, as planned, by 9 per cent in carrying traffic compared to 2004, with 0.5 per cent growth in volume and 10 per cent in value due a 90 per cent fuel surcharge mechanism. So SN's European operational performance, or fleet contribution, tripled compared to that in 2004.

This picture was not too bad, but nothing changed between estimate and final results. So Mr Davignon's observation during our last meting in October 2005 (see page 34) very relevantly underlined the nature of the challenge: 'Good news is one-off, bad news is recurrent.' SN's performance faced the same phenomenon.

The good news came from capacity management, which had saved more than €41 million with 9 per cent fewer flights. This result was healthy and the cutbacks were appropriate, because they were driven by research on the legitimate accessible market to each destination over the year. Consequently, this was cost effective and strategic too. In this context, SN's European network then had a clear idea of which rock-bottom perimeter had to be fiercely defended. But that did not say anything about the effectiveness and efficiency of SN's conquest of the market, which was a radically different topic not to be confused with maintenance (see *Organic Growth*). Finally, the approach developed over the previous 30 months avoided the company falling into the trap of marginal cost accounting, which would have been committing suicide in a sector where the marginal revenue has to be extracted from marginally very high performance.

Marginal revenue begins with the 29th passenger on any flight in Europe, which is not insignificant in terms of sales performance, and SN was very shortly to find this out the hard way. For instance, to find a use for an aircraft during its day stop in Marseille during the weekdays, a test was made on Marseille–Geneva, which was a route that no carrier operated at that time, the first quarter of 2004. This scheduled flight rarely reached 30 passengers and in three months the route burned through €250,000 of cash due to uncovered variable DoC. Many factors can explain the results, but the purpose of this example is just to remind us that business does not naturally follow just because a flight number is in the booking system. To hammer in the crucial importance of accuracy in cost accounting, I can repeat the comment of Jean-Marie Descarpentries at Bull: 'From now on we take any deal which is generating €1 contribution instead of this standard rate of 29 which means nothing to me. I want an order where clearly I can see the generation of margin and therefore I can negotiate accordingly.' This could be read as realistic, straightforward entrepreneurial behaviour – which it is. But one must note that it is also driven by a good dose of cost accounting common sense, which one must always retain. For SN the good news of the capacity-management measures could not be repeated every year, and the impact of that fact began to be felt in 2006.

Another piece of good news was that the unit passenger contribution increased by €12. This incorporated nearly 90 per cent of fuel surcharges. As of January 2006, the net impact of fuel surcharges 2004 on 2005 was just €2.5 difference, with a possible risk of reaching a ceiling in terms of what was acceptable to passengers. Consequently, this fantastic revenue boost free of any company creativity regarding price rises is currently vanishing from net performance.

The flat growth in number of passengers is, on the contrary, bad news. SN's passenger growth is in line with airport performance and outperforms the middling results of the Belgian market. Nevertheless, the strategic concern is real. A business with no volume increase is always problematic! What can we say about a shop that is not bringing in new customers? It is declining. Then the real question is whether one is not developing a form of Kenzo syndrome (see page 136), selling more and more expensive services to fewer and fewer passengers. Finally, SN's home airport is rocketing to a level of airport taxes that is not helping the perception of competitive prices for the airline operator. This is just a confirmation of the red lights that were already showing when the estimate of the year-end result and the forecast for 2006 were prepared in late August 2005.

The next question was then whether the new context would threaten the turnaround dynamic. After its very successful relaunch, SN was now entering a higher league of difficulties, where it had to struggle hopefully with a reconfirmed confidence in its own business model. The first question was what the consequences of such signals would be: tougher business conditions ahead.

BEAR IN MIND

A transformation mechanism is based on some form of creativity or at least new ideas, which can even concern management areas such as performance measurement. Their potential for transformation is influenced by a sequential mechanism. The different techniques for managing performance improvement have to be shared and sold in terms of their contribution to writing a new page of the corporate sustainable future. But this series of workshops was not a guarantee that the new management practices would be enforced.

Reaching this stage is a question of having a critical mass of educated users among the company's managers. This requires two things:

- Expressing the output of this transformation project in new corporate skills at dealing with a fresh information file, the route binder, structuring both the data and the analysis of performance appraisal at the elementary level of strategic execution.

- Providing a training platform to transform the skills into reflexes, the Route Business Strategy meeting or RBS. This RBS, or management forum, was a transformational arena where all the inhibitions that had been deeply rooted in the past disappeared, thanks to a formal process of rotating leadership among the various groups of stakeholders. The corporate doers performed many roles including active participants, preparation leaders and action plan controllers, in order to gain a deeper understanding of the current situation and become sharper in diagnosis and more committed to the action plan to be executed to improve the performance of the route. This process first supposed a high level of collective discipline to systematically repeat the same RBS meeting storyline under the leadership of a senior project owner; and secondly required a careful assessment of the progress achieved during these sessions, in order to recognize how much support was necessary and when these techniques would become management reflexes of best demonstrated corporate practices.

Over 18 months this goal was reached.

This kind of learning process leads to positive performance improvement results, and SN has unambiguously demonstrated that through the systematic enforcement of a new and sharper winning formula. The never-ending search for new sources of performance improvement erased one after another the ill-founded reasons for accepting mediocre performance, such as on the weekend southbound destinations. This in-depth collective iteration operated as a means to broaden the scope of analysis and to reconfirm the meaning of the strategic positioning, 'Connecting Brussels'. But these transformation techniques must be put into perspective through the question: could the organization hold to the pace imposed by the competitive context in which it had to operate? The beginning of the answer starts in segmenting the origins of performance improvement, sorting one-off measures from recurrent ones to read the future with more lucidity.

PRACTICAL REFLEXES TO SPEAK THE CORPORATE PULSE FLUENTLY

- Do you have a forum, a strategic dialogue platform, where the transformation of the company is scrupulously tested? Is its audience broad enough to create a critical mass of managers who are thinking and behaving differently?
- Which efforts of your staff show you that it is a best demonstrated corporate practice?
- Analysing the improvement of performance, how do you sort one-off measures from recurrent ones?
- How do you anticipate the next recovery ceiling? How did you treat the previous one? Are they representing merely an evolution in the difficulty of prolonging the current winning formula or a radical reengineering of the approach? Are you communicating enough both upwards and downwards regarding this evolution?
- Do you have a formulation of the meaning of a successful in-depth turnaround for your company in its sector? Is it shared upwards and downwards through the organization?

COMMENTS AND OBSERVATIONS

Arthur Lok Jack, Chairman of Neal &
Massy Holdings, Guardian Holdings and
Caribbean Airlines, Founder of
Associated Brands International

*M*any authors have already addressed similar business situations to the one Jean-Frédéric Mognetti faced at SN Brussels Airlines, but here there was a different solution to the challenge: developing a disruptive cost accounting system and culture to instil a sense of profitability throughout all the levels of the entire company. This corporate performance insight begins with in-depth questioning about whether one is measuring what makes sense in management terms to support a permanent performance improvement. This experience, which shows the transformation that can be achieved by substituting what is technically correct with what is managerially relevant, is very much in line with my own management philosophy, which I have kept fine-tuning for the last 35 years.

This book outlines a practical method for remaining focused on what I consider to be the foundation goal of any organization: developing the lowest possible cost structure, first to keep the widest range of strategic options available, secondly to organize a continual and structured dialogue on performance improvement.

The relevance of the management lessons of Parts IV and V of *Out of the Ashes* is not limited to the airline sector. Putting these chapters into perspective with my own view of business, I take for granted that genuine entrepreneurs know how to play differently with the word creativity to exceed the average performance of their own sector and to thrive. This statement could be perceived as mundane, but the emerging country of Trinidad and Tobago, in which I began my real entrepreneurial life 33 years ago at the age of 28, provided me with an unparalleled business education. From the beginning I was taught the hard way, next to the production line, that product affordability precedes value for money. If one doesn't assimilate this principle, one can have no sustainable business perspective. Affordability only knows one gospel: strict cost management discipline turned into a corporate management reflex and culture.

Affordability is not an either/or concept, especially under the pressure of globalization: the characteristics of one's own products must perfectly be on a level with their global brand equivalents. But the dominant flavour of my competitive formula is the lowest possible delivered costs to make my product accessible to the mass of my own country's population. Reaching the lowest possible costs is a real challenge when structurally one is operating with diseconomies of scale compared to global players. But paradoxically, one is helped by the global players' economies of scale in their own markets when affordability becomes the focus point. Economies of scale in one part of the world can become a constraint somewhere else. This does not represent a strategic barrier but a precious opportunity for developing a difference, which can become the root cause of one's own competitive advantage.

For instance, in the chocolate bar market, one company in my portfolio controls a very dominant share in the Caribbean with an international brand equivalent candy bar of 28 g sold for 100 c, while the international brand product offers a 52 g bar for 300 c. A MBA class discussion would conclude, borrowing Philip Kotler's words, 'I give more for less'. This is true, but it is merely incidental. The first driver is an affordable formula for mass consumption. Secondly, cost-management effectiveness makes this affordable product unaffordable for the global player in terms of the cost of access. To address this kind of niche at the global level, in a relatively high-value product easily shipped from a few manu-facturing premises around the world, a global player has to accept diseconomies of scale, which would weaken its own leadership in some other parts of the world with stiffer competitive conditions. Therefore in emerging markets its objective is to enjoy the high-end–lower-volume–premium-price segment. Good news – to a limited extent!

This does not mean that my company is not challenged. We have to face two kinds of pressures:

- First, other players coming from other emerging countries, from South America for instance. In this case it becomes a game in the same league, a race like in the Americas Cup with only one seat for the winner.
- Secondly, one's own customers are resegmenting their needs as a consequence of the emerging country's fast and sustainable GDP growth. This catch-up trend closes a part of the gap with the global player, providing a new category of customers ready to pay more for something slightly different, which is good. My constant challenge consists in staying with this evolution without making concessions, in order not to lose any volume.

This business context therefore demands that we are always sharper in terms of cost management to prevent the barrier being lowered against the global player. But the business picture also includes an attractive silver lining. Our affordability culture and its associated cost-management skills can also become a passport for welcome access to the western distribution world, which is constantly searching for more competitive procurement sources to support its own in-house brands. But realistically, one must simultaneously recognize the volatile status of incumbents in this sector, unless affordability has become the theme of a drastic in-house business education programme. My never-ending search to exceed the limits of our own competitiveness forces me to keep developing a programme similar to Professor Mognetti's at SN.

However, at the same time, affordability isn't a universal key that allows all the forms of 'organic growth' development to be attained. Emerging market conditions are not easily transferable to more developed countries in a greenfield way. For instance, the branded market is almost unaffordable for us in a straightforward organic growth approach. I would face a very high and costly barrier of brand recognition. Consumers already benefit from very solid brands that they know, trust and like. A solid unique selling point can make them try something different, but the brand power is so strong that there is no guarantee that next time they will repurchase from us rather than returning to their initial preference. Overcoming this barrier is virtually unaffordable for a brand from an emerging country.

Therefore, I need to add external growth skills to my organic growth ones. This means that when an emerging country player wants to enter into more developed markets, this has to be done by acquisition. Am I saying anything disruptive? Not at all, I am just referring to those who earlier found the key in combining low delivered costs with high perceived value. Seiko for instance, the Japanese watchmaker, tried to enter the high-end watch market controlled by the Swiss watchmakers 30 years ago. It failed until it found the opportunity to buy a member of the Swiss high-end watchmakers' league. This was Lassalle, not the most renowned company, but nevertheless a member of the club. The access to legitimacy was therefore possible and the package of skills with which Seiko had built its success so far merely demanded to be implemented with some adjustments along the way.

The lesson of this story is that one should not waste money uselessly in a game where someone else has already shown us the successful exit route. The famous playwright George Bernard Shaw wrote, 'We have learned by experience that men never learn by experience.' I cannot afford to make this kind of statement, because if it were the case I would have never built this group. To prevent falling into Shaw's trap, my recipe is to permanently remain on alert thanks to a continual search for knowledge and experience.

My observations about affordability and its associated cost management excellence are very much in line with what I read in these two parts of this book. Mognetti's affordability in the SN story takes the form of a unique indicator of corporate performance, a corporate pulse named RoDoC. In any enterprise that has the ambition to do something different in performance in its respective sector, one should naturally find an equivalent tool or concept to RoDoC or affordability. This raises a question for any CEO, board or chairman of an executive committee: what is your own corporate pulse? From experience, I consider that the answer is trickier than it would appear.

Mognetti's corporate pulse approach is disruptive and does not follow the mainstream or business fads. Consulting firms usually knock on my door with some form of 'corporate dashboard' that is technically correct. But I regularly observe that even though these are presented as tailor-made, they not only are closer to one-size-fits-all, but are also not relevant enough to support an in-depth corporate transformation and secure a healthy strategic future. The fallacy consists in believing that one is relying on the equivalent of RoDoC-affordability by focusing on the nitty-gritty aspects of one's own business. In reality, by doing so one misses the broad sense of direction that one needs. Even worse, nitty-gritty measures always demand a substantial dose of effort to reach their cruising speed. These efforts make one feel good and lead to the impression that one is delivering what is needed. But at the end of the day, the transformation or the competitive position-ing improvement is not radical enough to turn the page from the past. RoDoC and affordability are two powerful tools to avoid falling into the management trap of rearranging the chairs on the *Titanic*. They are corporate flags that focus the company on what is essential.

This book offers some practical lessons in terms of implemen-tation. Mognetti never operates with a 'from now on' style. He takes the risk of allowing two systems to coexist. This is daring. But his corporate pulse philosophy is driven by an insightful sense of prioritization, combining analysis and common sense, recogniz-ing through a remapping business model which are the most critical items; for instance direct operating costs in SN's case. This developed focus and rebuilt internal confidence in an organization that went through many years of poor performance and had con-sequently lost its sense of the correct priorities. Therefore the corporate pulse reinstils priorities. The approach described in *Out of the Ashes* is a means effectively and seamlessly to substitute the frames of reference necessary to make a new future happen and prevent the company returning to a crisis.

A corporate pulse is not enough to support a transformation. Affordability cannot come without cost management discipline. Both my own experience and that at SN report a way to play with cost accounting differently. The goal is to exceed the initial objective of reporting the cost of an operation, to assimilate a sense of profit building across the whole organization. In these two parts of the book Mognetti expresses his own corporate transformation proposal, where people operating in cross-functional teams are the irreplaceable driving force as long as they are supported by a correct, relevant and appropriate toolkit.

In this respect the book provides an answer to a difficulty I have encountered in cascading a sense of profit down the whole organization. This is critical, because it conditions a distinctive sense of reaction, which makes the difference between superiorly performing corporations and average ones. Mognetti's paradigm is simple: are those who have access to performance measurement able to understand its meaning and to organize the reaction to it or to push harder for success? He proposes a transformation process that is achieved not through a superiorly brilliant tool, but through an accessible, even obvious concept. Therefore, in his approach the effort shifts from understanding the approach to assimilating it or playing with it. The book operates as a reminder about creating enough opportunities to play with a new concept that have an increasing level of difficulty. This represents a nuance that I did not enforce as systematically as his experience demonstrates and recommends.

In many businesses in my portfolio of companies I have often explained to my teams that I consider cost accounting as the equivalent to a floor plan, while profit is the elevation. Unfortunately, my experience tells me that this translation of floor plan into elevation does not occur by itself as one would expect. Managers too often remain stuck to the technical side of a management tool. They mistake the trees for the forest. Therefore, I read Mognetti's sense of a profit proposal as a management development process for systematically bridging floor plan and elevation, whatever sector one's business is in.

Moving from floor plan to elevation is primarily a question of training; more accurately of perfect training. That is what is addressed in SN's forum at the business unit level, or RBS meeting, where the goal is to enforce discipline in the quality of the performance enhancement dialogue between all stakeholders around the deviation analysis. This approach, which I use with some minor variations, represents the best repellent against corporate complacency or routine, which imperceptibly blunts one's competitive sharpness. I keep repeating to my troops that I want to hear the story behind these figures – massage the figures and they will tell you the truth. I trust in the discipline of deviation analysis, which helps one to reach the root cause of one's performance. This demands that one is systematic both in the preparation and in the dialogue. Therefore it creates a context where solutions become more natural, more spontaneous and more numerous. One also builds the possibility to keep choosing between alternative options – a characteristic of solid corporate health.

Finally, *Out of the Ashes* offers us a method to assess when the profit reflex has spread across the whole company. The story of the RoDoC ten commandments is illustrative of a genuine signal of transformation. I agree with Mognetti: when the team can take the time to have fun with their own management concept, this means that one does not have to hammer it into them. It is in their genes and the cruising speed can increase.

From floor plan to elevation or from cost accounting to profit requires a critical dose of Mognetti's educated creativity, which does not occur by accident. A disruptive concept to focus energy and attention on the true priorities of a business model combined with a well-oiled dialogue tool explains, in my opinion, a substantial part of the hidden reasons for the superior performance of the most successful companies. Through his experience and concept, Jean-Frédéric Mognetti draws the lines on the pitch to increase one's chance of success.

PART VI

SEEING THE FUTURE FIRST

*T*he recovery process implemented after February 2002 at SN was driven by a key ambition, to turn the page from the past in order to invent a radically different company, especially in relation to performance and culture. The company was heading towards this goal thanks to the efforts employed in both building a new context and developing new management tools. This was a relatively short-term game, but had the benefit of not compromising the relaunch process. In this respect SN was doing quite well, but without any delay a higher league of challenges needed to be addressed, to deploy these skills and techniques over a longer period to gain more control over the company's future.

The future is especially critical in a sector such as airlines. By one's own capacity to anticipate the future, one gains control over the present where there is no second chance of correcting performance. In SN's game there is no possibility of building inventory. Therefore the future must be addressed with a lot of care, both in planning the improvement of one's own quality of performance and in developing means to diminish the risk of seeing one's future compromised by outside events. Consequently, as its next challenge, SN had to jump into the management of uncertainty. In the philosophy of forecasting I suggest, I do not promise to get rid of uncertainty, as too many authors do, but to deal with it more easily and consequently more comfortably, transforming it into a characteristic instead of a constraint.

Forecasting is a tool that supports execution. My objective was to design this tool for business doers, these who are jointly accountable for performance. But simultaneously, the tool must also invite the executive committee members or senior managers into the day-to-day management of the business unit's performance, in order to confirm a direction or be warned of some concerns. To reach this result, a forecasting system never exists independently. A correct management context must precede it. Forecasting as a technique sets a frame of reference, which delivers disruptive support both to the business unit management and the executive committee about whether, first, 'nearly in real time' performance measurement is achievable; and, secondly, deviation analysis has become a reflex at the business unit level. These were two of SN's characteristics, which allowed it to move ahead into forecasting in mid-2004.

The last part of this management turnaround journey goes through the different steps that led to the implementation of this management tool, underlining its impact both at elementary level of strategic execution and corporate level. First, plugging an organization into its future goes far beyond a budget dialogue and process. Moreover, forecasting is the symptom of an organization that is at ease with itself, but this situation does not prolong itself by accident. In the specific development of the different steps of the tool, the risk of rejection has to be managed from the pilot phase of the implementation. Finally, based on the real facts of the SN case, the tool shows how the management of forecasting at the business unit level also turns into a strategic warning system at the corporate one. Based on the seriousness of the warning, this may lead to a restructuring plan being proposed to prolong the success of the relaunch and prevent the turnaround becoming no more than a glorious remission.

CHAPTER 13

FROM BUDGETING TO FORECASTING

BUDGETING: NOT AN APPROPRIATE TOOL FOR DEVELOPING A FUTURE-CENTRIC DIALOGUE

Forecasting does not result from the extrapolation of a budget process. Even in a newly reborn company, a budget can suffer from the same endemic weakness that I explained in *Organic Growth*, the risk of arm-twisting. Forecasting intends to develop more maturity in the relationship with one's own future performance, which allows one to cope with this ambition to see the future first.

As of mid-2004, SN's management effectiveness process was delivering some tangible results. Performance measurement had been reengineered and enforced at the elementary level of strategy execution thanks to the strategic dialogue platform. Two managers were jointly accountable for the performance of each business unit, routes in Europe and gateways in Africa. The strategic dialogue through the RBS system was operating very effectively. In the management development process, I had very early on – as of fall 2002 based on the experience of preparing the 2003 budget – reached the conclusion that this company should become more mature and as soon as possible address its future. This would not be a painful exercise like the 2003 budget, but a natural consequence of the organization's in-depth understanding of both its business environment and its own performance.

Business story 13.1: 'Now let's go for a beer!'

It was clear that we were miles away from this objective. I made a marginal contribution to the 2003 budget process. One evening in fall 2002, almost by accident, I was sharing with the Sales VPs of Benelux and Europe what we were trying to develop in terms of route-performance assessment with the HEC MBA participants consulting project team. The MBA team had estimated revenue goals for 2003 with a radically different approach. I proposed that the three VPs should compare their revenue goal against the route BEP that we were beginning to assess. The revenue evolution was a first point of comparison, but translating it into something more tangible seemed to me even more illustrative of the challenge.

My suggestion for 2003 was, for instance, to estimate which volume development – number of pax not euros – was needed to reach BEP. At this stage of the company's evolution, the situation was highly influenced by a context combining low-cost airline pressure on one side and SABENA's past performance on some specific routes on the other. So the sales team's legitimate hypothesis was that yield could only diminish, which was quite correct based on the observed trends. Consequently, the challenge was a question of pax growth in the majority of cases to maintain or exceed the previous year's revenue. The idea of costs had not yet got onto the sales department's agenda, with its associated dynamic of profitability building. The results in terms of passenger growth were the first level of challenge; once these figures were translated into pax per flight the passenger numbers development challenge became quite frightening and in some cases virtually unachievable. Pax growth was a consequence of revenue increase, which was highly influenced by the Executive Chairman.

Having introduced this 'physical perspective' on the budget target, then performance could be observed in terms of route break-even gap, for instance how many passengers per flight below the break-even or above did the targeted performance represent. Such a small trick changes the vision. In fact, this was Keating's trick in *The Dead Poets Society* when he asked his students to sit at the professor's desk to change their vision of the environment. It was my first real contact with the sales team to instil the importance of the cost dimension in how the sales mission was visualized. In their efforts to define the next level of performance, nobody had yet addressed it from the costs side. This meant that the appropriate level of service, the frequency per week for this specific route, was taken for granted. My simple question broke a paradigm in asking incidentally, can we afford this level of service? My hidden goal was to kill two alibis, first, 'the customer demands it', which could be true; and secondly, 'this is strategic'. Paradoxically, I note that losses are always strategic!

My questions suggested that they should gauge the legitimate accessible market and consequently accurately adjust asset deployment accordingly. Simple but not so easy, because one bumps into a deeply rooted cultural practice. Salespeople endemically develop the reflex to put the asset first and then run after break-even, which usually is never reached. But simultaneously, I have observed a costly senior management fallacy, which consists in saying that one must not inform the salespeople of the costs because they will therefore limit their ambitions. Both are management nonsense.

The team was focused on finding new traffic, which can be appropriate as long as it is validated by solid facts. But putting too much investment into a market characterized by an existing overcapacity is seriously hampering one's future. And this is made worse both by the potential acceptance by the market and the reaction of the competition. Market acceptance of the newcomer will certainly be high among a volatile clientele, but the most rewarding position, which one always has in the back of one's mind as a conquest target, will be fiercely defended by all means by the incumbent. Therefore this technique of development also raises the question of the company's skills and track record at achieving real conquests. In contrast, adjusting the level of asset deployment, which directly generates some savings, is a more conservative approach that avoids putting the cart before the horse. It depends on people's willingness and courage to make it happen, while revenue-based measures are influenced by the business environment, which is far more random.

After this initial discussion (which in fact announced the structure of the future Route Business Strategy meetings implemented a year later), I saw the three VPs concerned about their growth targets, unfortunately already accepted for the vast majority of them, with the help of pressure from the Executive Chairman. And the most senior of them concluded, 'Now let's go for a beer!'

In fact the VP in Business story 13.1 was right, we all needed a drink to anticipate what followed this meeting. At the next budget session that was supposed to end the process, the three commercial VPs were victims of a superb arm-twisting display from the Executive Chairman, who wanted to present to the board a budget with particular results. A few months down the road, becoming more intimate with the business reality of the network, I noted that on some routes the results of the arm-twisting exercise were far beyond the accepted objective and exceeded in some cases the size of the market. Even in a just-reborn company, DHL's bias that I described in *Organic Growth* regarding the fight between the outstations' management and the headquarters' functional staff for performance forecasts was already in place. The lesson was simple: the traumatized survivors of the SABENA disaster were ready to accept anything, because their own confidence in the future of the company had not yet been rebuilt by their own accountable contribution. So they accepted whatever the senior management – the CEO and Executive Chairman – wanted.

> **Stretched goals in a just reborn company are not cosmetic – they can hurt badly.**

This kind of situation is made even worse when the CEO and the Executive Chairman have a limited, day-to-day, in-depth experience in the business. Consequently, at this early stage a CEO must be very prudent in what they require from their troops. They are not yet committed, but they judge the appropriateness of the pressure and it is very difficult to change this first opinion. In this specific case, Rob Kuijper's image did not come out of this story unscathed. It is never a case of pushing but of leading, especially at this stage in the life of the company. So, I decided that among my management skills development goals, a tool and the associated attitudes to transform this deep dive into the future into a more natural process should quickly become one of my priorities.

FORECASTING: A SYMPTOM OF GENUINE CORPORATE HEALTH

'Seeing the future first' can be perceived as an inappropriate ambition, especially in a company like SN where the next week could already be in question if one purposely takes a worst-case scenario. In my opinion it is not inappropriate, however, for two reasons:

• It is a legitimate ambition for any good management team.
• When an organization can achieve it, this is a symptom that it has successfully reached another level of performance. This result means that a lot of other management areas are now under control with the appropriate dynamic. This observation also works in reverse. An organization that senses its first signal of bad habits usually starts to question the rationale of the tools that are plugging it into its future.

Decreasing the frequency of the YEP report, as explained in Business story 13.2, was a simple question of resources. The superb team who were managing this report, focused on speed of reaction and a short-term future, had lost its number one customer when

---

Business story 13.2: 'From now on only once a week'

I have maintained a good relationship with many SN executives, with whom I regularly stay in contact either over the phone or through face-to-face discussions over a drink when I am in Brussels. In May 2006, one of them sent me an e-mail to discuss reporting principles. During this conversation, he informed me that the YEP report (see page 141), which was available at 11 a.m. each working day on the desk of the executive committee members and some senior managers, is now only circulated weekly.

This is an example of how to become 'penny wise, pound foolish' organizationally speaking. The justification was very simple: nobody looks at it every day! That is an excuse for a decision based on complacency. Will people look more seriously at this report if it comes out weekly? Then why keep it to weekly? The next step could be a monthly circulation. But if that is the case, then it will no longer be necessary to produce the report because it will be circulated something like 20 days before the final accounting-based route performance report is available. So, if one wants to rationalize, the intermediate steps are mediocre decisions. One must have the courage to say that something is irrelevant, we don't trust this approach so we must stop using it. However, the rationale for this decision was far less strategic.

Peter Davies left. As a CEO he was oriented to 'seeing the future first', and this management characteristic was clearly observable in his style. Therefore after he left:

- There was no sponsor to nurture the practical enforcement of the concept in the company's day-to-day approach.
- The team also saw his direct boss changing his professional way of life by applying for an 80 per cent, part-time contract.
- Finally, the manager of the monthly performance reports was encountering a resources issue.

This led to a decision to merge the two teams to get some synergies in resources. The head of the newly formed team saw the arrival of the additional resources as a way to help him produce a majority of past-oriented documents, which incidentally were also critical for the company. In this context the priority is naturally reshuffled and the value of a systematically future-oriented approach was not his key focus as in the previous organisation. He couldn't be blamed for this, but the senior management could, because this apparently insignificant decision, taken in good faith, in fact unplugged the organization from its future-building mechanism. So this lack of organizational insight just diminished the company's autonomy of reaction by damaging its capacity for anticipation. The lesson is simple: dealing with the future is a very fragile topic that demands a lot of care to, first, make it stay alive and, secondly, produce improvements in organizational effectiveness.

> **The future is a fragile organizational topic surrounded by a lot of predators.**

ASSIMILATING THE LESSONS OF THE PAST IS A PREREQUISITE

Once again, the context is one of the most critical elements. And the question is also simple: how can a CEO ask the team to look beyond its view range when the bankruptcy had made its own benchmarks almost meaningless? As Jack Welch said, 'To be a good boss one needs to be well with oneself.' The situation is the same, because a consistent future is based on a well-assimilated past, either in prolonging it or freeing the organization from it. The CEO must make the choice quickly. He or she cannot let things remain unsettled and at the same time nurture the chances of success of the turnaround. Consequently, it is unrealistic to expect tangible progress as long as there is no solid technique for interpreting past performance. This is critical in the case of a turnaround in order to prevent repetition of the past's mistakes. For a healthy company the issue is slightly different, and consists in preventing yesterday's success becoming today's or tomorrow's failure.

> **Future clues are dictated by the keys of past performance.**

'Seeing the future first' results in the possibility of developing a forecasting approach, tools and attitude dialogue at the elementary level of strategic execution. It requires that the confidence-rebuilding process has already delivered its results. In this new context, the company is back on a healthy growth track because the management can cast itself into the future without any apprehension. This means that past results are no longer a catalogue of excuses about why the company performed so poorly or a wimpy discussion on why the good results will be prolonged. So emotion has disappeared and forecasting is just a consequence of an advanced understanding of one's current business at the elementary level of strategy execution.

But even though forecasting is a legitimate goal in a turn-around process, some serious constraints have to be recognized. It cannot become a new topic for management skills development, because the company is already very much occupied in organizing its own survival and swamped by its daily business constraints. Consequently, the risk of rejection is very high. So, the trick is to develop these skills by imperceptibly getting the company's management to use them and the associated mindset. In summary, it is a way to save time and to keep aiming at the long-term objective of the turnaround by leveraging the company's management skills.

FOCUSING ON THE UNCERTAINTY SIDE OF PERFORMANCE BUILDING

After the implementation of the new performance measurement metrics, YEP and RoDoC, the development of a forecasting tool quickly ranked very high among my priorities. With his experience of DHL, a superiorly well-managed company at least until the late 1990s, Peter had at the back of his mind the idea of relying on what he called a rolling forecast system, which worked quite well at DHL. But the weaker aspect of the DHL system was not the costs side, but the revenue side. This was the reason for my first contact with him 12 years previously, when he asked me to develop a means to improve it. This concern remains quite hot in the company. When I was gathering the different perspectives I wanted to synthesize in this part of the book, I contacted Doug West, who was until early 2006 DHL Commercial Director Europe and previously responsible for European emerging markets. His e-mail on the revenue side of forecasting was quite illustrative of the nature of this challenge:

> The issue is still the same with the new company, it is just the scale of the problem which has changed . . .

> **Revenue forecasting is not yet satisfactory. Unfortunately the scale of the challenge has also dramatically changed.**

On the costs side, the forecast is handled mainly by templates and the proposed output is fairly secure. On the revenue side far more is required to reach the same level of certainty. Different business conditions must be considered, such as intensity of competition, rate of growth, number and size of competitors, market concentration and much more, in order to grasp the overall reality. That is why the RBS dialogue addressed in the previous part is a form of training pitch to deal with contingency to anticipate the future-forecasting exercise. This leads to a business attitude where the outcome is less important than the process itself, because its goal is to reduce uncertainty and not to get rid of it nor to master it. I developed this approach a long time ago under the influence of Professsor

Andre Tunc of the University of Panthéon Sorbonnes in Paris, my thesis director, who systematically repeated during our progress review sessions: 'Mognetti, one can live and keep progressing with well-defined unanswered questions, while a wrong answer is the insurance of a cul-de-sac.'

DEVELOPING A FORECASTING DRIVING RANGE

SN's forecasting tools were developed in a form of 'skunkworks' in order to prevent the feeling of an overwhelming workload among senior executives. The overall frame for the forecasting process was ready quickly, but once again it was a question of how the team of senior managers would accept it and modify their own practices in order to really embrace the uncertainty of the business with a different mindset.

Whatever approach is used for a budgeting process, seven or eight months before the end of the year the key question is what estimation of volume of passengers, revenue and consequently margin can be achieved, so that the accountable managers commit themselves to these figures. This idea is not just to get a figure, but to review through an in-depth intellectual process at the elementary level of strategy execution the real performance-building situation of the route, to prepare the first step of 'seeing the future first'.

To do this, my suggestion was to run a workshop that was as light as possible in terms of preparation for the managers involved. This exercise was run in teams to test the capacity of joint route managers to explain the rationale for the future performance of their respective routes. Some interactive support had to be designed to play with the leverage points of performance generation. These leverage points in the SN case were in relation to revenue or offers. This meant, for instance, that the offer on a destination was expressed by its frequency, which could be diminished or increased with an automatic result on the business unit's BEP. But simultaneously, this led to the question of the combined capacity both of the brand and the sales team to concentrate or attract traffic, and consequently positively to leverage the decision. To achieve some tangible results, this kind of support tool helped managers become lucid about the results to be achieved.

> **An effective forecasting technique begins with the end-of-year estimate.**

Such an approach allowed managers first to gauge the sales team's view of performance, and secondly to pull down the walls between the various stakeholders at the elementary level of strategic execution; in this case network, revenue management pricing, sales and marketing. When this workshop was run the Route Business Strategy meeting did not yet exist, but because of the audience involved this form of organizational evolution was already being foreseen. Finally, to prolong the impact of this workshop and to create a common basis for performance assessment reflexes, I prepared for each route manager a summary by route with the same set of critical information in order to generate questions regarding performance forecasting. But this was also a means to bring about a commitment to make it happen, because it expressed our collective best assessment of the business situation. If we could not make it happen, had we missed something that we could deal with next time?

Summaries such as that in Table 13.1 resulted from the discussion of each route team's figures. When I sent the participants the written results, not many were received without any additional discussion over the phone. They needed to reassure themselves that this was really what they were accountable for.

Now we had to move to the next step, developing a formal tool to cast into the future and to operate beyond our conventional view range to cope with the time span of the sector's evolution.

Table 13.1 Example on a specific route.

Scenario III, stronger cutbacks during the weekend and no 319 flights

End of the year

- 158 403 pax +6% pax
- Average pax contribution €157
- Number of flights 3117, −6%

Profit €9925 million, +10% compared to 2003's workshop estimation

PILOTING THE FORECASTING APPROACH

The pilot phase aimed at delivering a plug-and-play product. Its goal was to familiarize management with its use. Even though we took a lot of care developing this tool as a logical development of what had already been implemented in order to avoid the risk of a new toy every day, the launch of the forecasting approach still held a surprise.

Which timespan?

Forecasting is a means to mark out the path, but the question is how much of the future has to be covered. The challenge is to leave aside the mandatory aspect of the budget in order to address the future with the appropriate management perspective. In April 2006, I had a discussion with the CEO Europe of Rieter Automotive, who explained that for the company's mid-term strategic process, five years fitted the automotive industry business context when one considered the lifespan of an automotive model. Moreover, the budget perspective was hardly a concern because the results had a probability of occurrence close to 100 per cent. In fact, the real challenge was to forecast the results two and three years down the road, where the probability of correct estimation fell respectively to 90 and 80 per cent. This cumulative 30 per cent represented the level of uncertainty that this CEO would like to gain better mastery of so that the managers at the business unit level could reach the appropriate level of confidence to keep generating above-average performance. Thus in Rieter's case, the idea was a forecasting approach over three years. In SN's case it was a little less, three to four IATA seasons, so 15 to 20 months down the road was an appropriate range for changing the paradigm of the company.

In my opinion, the forecasting range is something that is ad hoc, which needs to be reassessed on an ad hoc basis. In a business with a high level of uncertainty, I suggest reducing the scope of the forecasting period to something that is in line with the company's current operating challenges. In 2002, I helped the CEO of Kenzo (see page 136) prepare his three-year strategic plan. This document was a compulsory milestone imposed by the group. Retrospectively, I consider that the business situation of the brand dictated three different subpositionings: Paris, Jungle, Jeans. In its different distribution channels the forecasting should include three collections. Consequently, if I had had time to develop this tool in this environment, I would have opted for a forecasting timespan of 18 months to cover the intertwined nature of the company's performance. This would have allowed managers to relearn how to manage commercial success with the appropriate attention, which is not 100 per cent related to the fashion value of a collection. In summary, each company must develop a forecasting timespan that results both from the nature of its business model and the management challenges that it is currently embracing.

A first draft

To stick to the objective of coming up with this management development tool, the first version of SN's forecasting document also took the form of a skunkworks. It would have been a waste to mobilize the company's resources on a tool for which its staff were not ready, even though it could be a paramount requirement. But the forecasting document did not land from the moon. In fact, it had a very important ally in the whole SN management development system. The forecasting document could be presented as a by-product of the discussions that had already been run between all the stakeholders in a business unit's performance during the RBS sessions.

The first real corporate forecasting document was prepared for 2005 with no involvement by the route managers. They were just invited at the end to discuss it as a reference-training tool to develop these new skills. The acceptance of this document required management of an unexpected risk of rejection, which I did not initially imagine.

Business story 13.3: Thunderstorm at 6 p.m.

The forecasting document was developed over fall 2004, as a dividend of the Route Business Strategy (RBS) discussion. Each business unit was the object of a specific forecasting section, made up of a formal synthesis in the form of a reminder of the market context and SN's relative performance, leading to a strategic diagnosis and some points for action. After this overview was a very detailed quantitative document giving the results of the forecast for the route. This was bound in a 5-cm-thick booklet and it represented the draft of the first version of the forecast. Ten copies were printed and circulated among the company's key stakeholders.

There was no reaction for a couple of days, until one evening at 6 p.m. I recognized the trap. In the office of the VP Sales Europe, I saw the VP Sales Benelux and the VP Revenue Management and Pricing. I walked in their direction

and from the body language of the VP Sales Europe I anticipated the problem. 'What are you trying to impose on us? This
will exceed what we have in the budget,' said Etienne, VP Sales
Benelux at that time.

Was that the issue? The budget was their reference document, but here I was in front of a real management development challenge. Forecasting meant nothing to them, it was not
in their frame of reference. How could we make them add to
the budget a different management approach, which also looks
at the future but is the one they need to focus on? It took me
a long time to explain simply that it was just a draft, a working
document, something they were not accountable for. The
reader can easily imagine the context of this discussion, with
people's eyes difficult to catch and a lot of words around the
key issue. I stopped this discussion when I suggested organizing
individual telephone conferences with all the stakeholders
in the route's performance to validate what was in this
document.

Their concern was simple. I was observing performance at
the business unit level based on the discussions that we had all
had during the RBS meetings. So the information was never
the issue. Moreover, the document did not take into account
some destinations that had just been launched, so that was not
an issue either. But the three VPs did not look at the pedagogical objective of the tool at the route level, but at the consolidation sheet on the second page of the document, which was
virtually a consolidation of the European network.

Business story 13.1 ('Now let's go for a beer') was without
any doubt reaping some of its dividends. My relationship with
the whole team was excellent, but we could always wonder
whether the Executive Chairman had found a new version of
the arm-twisting budget. The consolidation sheet created a real
tsunami, because they had agreed after the different budget
shuttles on a RoDoC of around 106 for this network and the

document was proposing a RoDoC of 111, representing a couple of additional millions of fleet contribution to generate. That was a real concern – more precisely a tangible threat – and frankly it was understandable. The budget had just been accepted, and in a turnaround phase a document that says +5 per cent was a shock.

The test took place at the end of November 2004, in the form of three days of telephone conferences with all the stakeholders, from 8 in the morning to 6 in the evening. This was done with the VP Sales Europe and those jointly accountable for each route in the office, with the country manager at the end of the telephone line. Each country manager received their own set of documents before the conference call so that they could prepare for the discussion. The process went very well and, for destination after destination, I could observe that the pace at which each route was validated was increasing because the confidence in the data was increasing. The data were becoming familiar and the team grew more and more accustomed to the process behind each forecasting sheet. Finally, one must not underestimate the positive influence of the horizontal dialogue between the countries after the first day.

All the corrections discussed during the conference call needed to be integrated into the forecasting document. But this came with a surprise. This exercise started with the idea in the back of each VP's mind that it would be a fair way to close the gap with the reality of the budget. The consolidation of the corrections increased the RoDoC by 2 points to reach 113. In the end we kept 111, because the goal was not the two additional points but the discussion, which we had held for nearly three days. This was a management transformation breakthrough, because the entire company, even in remote locations, was able to come to a common conclusion regarding a fair assessment of the appropriate ambition at the elementary level of strategy execution.

The context was difficult, where what was so nearly a rejection was transformed into a level of acceptance that I did not expect when I made the first copies of the forecasting document. That became a reference point for 2005. Once this step had been passed, the head of revenue management for Europe took the initiative to transform the document into a database that could produce tables, which he could link with the revenue management system to observe whether each flight was in line with the objectives discussed in the document. It is interesting to observe that once this level of acceptance was reached, the organization found a greater management impact for the tool, using it with a more integrated management perspective.

Beyond the management of rejection, the forecasting tool was also a fantastic opportunity to manage the fundamentals of the approach, to hammer in again SN's specific context and to build profitability improvement.

Forecasting: a means to consolidate performance-building confidence

The development of the tool included a support guide to provide managers with a reference frame, to remind them of the key steps of performance management and acquisition in the sector. The weight of asset costs in performance was so critical that the definition of the necessary level of service for a route to conquer what SN assessed as its legitimate accessible market was the systematic starting step. Secondly, I insisted in the previous sections that the management philosophy developed was based on the relationship between empowerment and accountability. This is inherent to the forecasting process. After their deep involvement in the RBS process, the route joint managers had to assess their commonly shared and accepted objective of RoDoC on the route. It was the combined expression of their common in-depth understanding of the competitive context and of SN's performance potential. Once this goal of RoDoC was set up, a second objective concerned the estimate of average unit pax contribution. These two objectives could then be transformed into a volume of pax that has to be assessed against the legitimate accessible market to answer the question of whether it was feasible, wise and ambitious enough.

The tool was built to extend what was discussed during the strategic dialogue meetings at the route level. The tool pursues an additional step: providing more management accuracy in the interpretation of forecasted performance. During the RBS workshops I observed that in the discussion we did not reach the root cause of performance, running the analysis with a level of information consolidation that was not sufficiently focused on the key issue. In the forecasting approach, I was obsessed by reintroducing a systematic analytical process that not only clarified the business reality, but also prepared the future structure of the analysis of the route business strategic meetings. With this document the goal was first to allow the accountable managers to commit themselves to their exact roles, and secondly to propose the appropriate reaction without any delay.

For instance, the RoDoC at a route level, due to the specifics of SN as a business-related traffic carrier (see Figure 11.7 page 224 showing the difference between weekday and weekend performance), was in fact the consolidation of two RoDoCs, a weekday RoDoC that was generally high and a weekend RoDoC below 100, which could dive to a very concerning level of performance. If the document was not able to deliver this nuance, an opportunity for appropriate reactions would remain hidden. Therefore the in-depth improvement of the health of the company would not progress. These two RoDoCs followed a historical seasonality, which also had to be reflected to keep track of the pattern of the business. Then the game turned into reconciliation, gauging whether the pax performance by generic booking class (Business, Time Sensitive, Price Sensitive) was a realistic goal.

This forecasting exercise also had to be able to dive into the details of the operation, for instance observing pax generation by booking class by point of sale, without confusing analysis and management goal. The best is the enemy of the good. This applies also to the management details. For instance, we saw in Part V that we could observe performance by day of the week, but in this exercise it would not be relevant in management terms to forecast by day of the week combined with seasonality. We could observe the consequences of the forecasting approach in this respect and reconcile the results with our own experience and the expertise of the business unit. This was a sound management practice, which avoided becoming lured into a jungle of details and losing the commitment of the team.

> **Forecasting selects the relevant level of detail for sound management decisions and follow-up.**

DEVIATION ANALYSIS IN REAL TIME

Once the tool was in a condition to be put into practice, it had to be implemented in order to nurture the management progress it foresaw. Once again, in this phase of turning around management skills, managers did not spontaneously enforce the tools in the expected new way just because they were available. Once the first monthly results were achieved, the goal was to keep delivering the analysis with the same tempo. The route YEP performance report was ready on the second day of the next month and the first task was immediately to deliver two deviation analyses:

- Year on year, a classic perspective.
- Achieved against forecast.

I prepared the deviation analysis for each route to be consistent with the way I introduced this hassle-free new tool. The goal was to deal with the management benefits of the end product. Once the deviation analysis had been performed, the information-giving task was over and a final observation could be made. Performance could be in line with forecast or not – that was merely objective fact – but more critical were the reasons for that particular level of performance. Introducing this discipline of recognizing whether or not the achieved result had followed the forecast was the expression of a day-to-day management that was commanding its own business model with insight.

The deviation analysis was influenced by the flavour of the short-term concerns that the company was currently facing. In this case, pressing events in early 2005 were the management of the cost of fuel and the associated fuel surcharge, which incidentally was supposed to cover the extra costs of operation. Figure 13.1 depicts this situation and simultaneously shows for one example

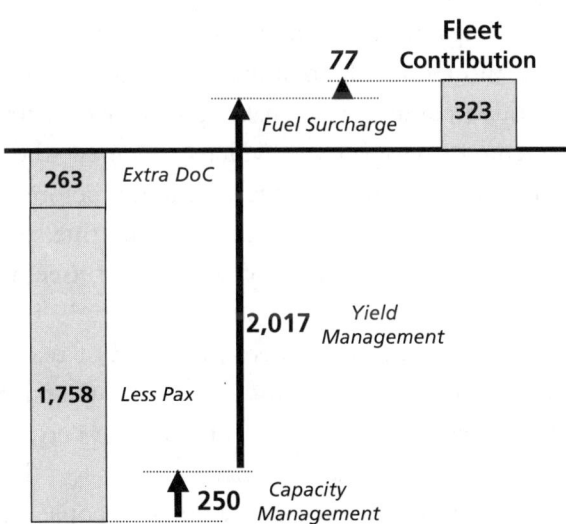

Figure 13.1 SN Brussels Airlines deviation analysis, forecasted vs actual, 2005 6 months.

how the fuel surcharge was covering the extra costs of operations, but simultaneously how the three other leverage points of performance improvement were affected.

Forecasting is a management technique and it always runs a big risk of confusion between objectives. The purpose of deviation analysis is not to show that the forecast is correct, but that the hypothesis behind the forecast works. To illustrate the philosophy of the approach, I found a brilliant analogy in a comment regarding the bicentenary of the battle of Austerlitz, which was a jewel of military strategy. French historian André Castelot wrote that in this battle Napoleon demonstrated his superiority to all the generals and strategists of the coalition in the way he dealt 'nearly in real time' with all the events of the battle, leading to one of his most memorable victories. But Castelot added that 'unfortunately Napoleon wanted it to be written that he had planned everything'.

The forecasting tool was not designed to develop a superior forecaster. It was a means to keep thinking about one's own business environment in more detail and to help develop the manager's speed of reaction. But once it was fully in place, it became a solid organizational competitive advantage, one of the stealthiest and most difficult to imitate. It should put SN into a better competitive condition by enabling it to react with a sharper micro-management perspective, which answered the question: What can I do in my own area of responsibility before it becomes a corporate issue? To stay with the military doctrine, the tool was developed to allow the attitude of a strategist (see page 202). General Von Molkte promoted a strategic attitude not at headquarters but on the front line, where the speed of reaction has its most meaningful impact, where the events occur and where the reaction can be immediately decided, because the person in charge is empowered and simultaneously accountable for their results. This is influenced by whether correctly trained people are empowered and in the correct place. Forecasting secures the answer to the first part of the proposal, while the CEO's management practices achieve the second part.

CHAPTER 14

PLUGGING DAY-TO-DAY PERFORMANCE INTO THE FUTURE

FROM FORECAST TO END-OF-YEAR FINAL ESTIMATE

The pilot phase was behind us and successfully overcome. The new tool had become part of the toolbox as standard practice for the company. Its value for the business unit manager was quite straightforward in delivering a means for continually questioning the alignment of their performance and the reasons for it according to what had been anticipated. But at this stage, I wanted to insist on some additional dividends that were more strategically flavoured.

At the end of the first quarter 2005, the forecasting system was in place and already used by all those at the RBS meeting. Over spring 2005, all RBS preparation and analysis was performed with the forecast as reference. The details of the preparation deserve to be noted. The tool was used by the route ad hoc team, including the managers of both ends of the route and some colleagues on the support side, network pricing revenue management. For this specific case, the rolling preparation was under the responsibility of the Benelux route managers, whose analysis combined the strategic perspective of past results and an estimate of the end-year results.

The estimated results had been also observed at the end of each month thanks to the 'nearly in real time' performance report. A question could then have been raised about whether it was necessary to reforecast the result in the light of the different information accumulated. This might seem semantic, but I did not want to transform the business unit managers into administrative clerks in front of a computer. My philosophy was to keep the forecast as it was. At a stage of the life of the company, it expressed what we had collectively thought the future would look like. The future would never look exactly as we thought, but we needed to keep a reference to start the next exercise with the possibility of continually drawing lessons from past practices.

Consequently, once we had enough months of operation under the existing forecasting system, a complementary tool was introduced to address this request for reassessment: 'final estimate of end-of-year performance'. The route team permanently maintained two databases, 'forecast' and 'final estimate of end of year', which combined in specific months:

- Final accounting results up to two months before the current month.
- On-time performance results estimated for the previous two months.
- Estimated performance for the remaining months until the end of the year, which was either the forecast or a reassessment if necessary according to the observed trends.

This end-of-year final estimate represented the tangible synthesis of the tools developed so far and their associated management philosophy – 'nearly in real time performance measurement' – to help anticipate the future result.

Again sticking to my pedagogical objectives and principles, I took the option of delivering to each route management team the estimate of their respective end-of-year result as an end-product. At this stage, it was not necessary to involve the route management in the reassessment itself, but the team was involved in the Route Business Strategy (RBS) meeting in an analysis of the results of this reassessment. In each case, this led to the implementation of some ad hoc measures, such as weekend flight cutbacks, to secure the route's performance improvement objective. Within a year, or two RBS meeting sessions down the road, the objective was to give responsibility for the reassessment to the route management team, as an integrated part of the RBS meeting preparation. In this context, it would become a simple reflex where one did not even note the effort to achieve it, because the mind was trained to immediately read things in the context of performance improvement leverage points.

Business story 14.1: One swallow does not make a summer

The memo analysing the synthesis of current performance on the portfolio of routes after the first quarter was entitled, 'One swallow does not make a summer, nor a quarter a year'. The title sounded a warning message for SN's management. Things were quite clear: the results were in line with the forecast. This was good news! But forecasting is characterized by a constant devil's advocate role. As developed in this book, it more often promotes the greater importance of the question over the answer. How the forecast results agreed with the actual results was not the key point. What was crucial was to wonder whether the result used the forecast leverage points as anticipated or whether the business model was sending different signals, which needed to be quickly analysed in terms of impact on future performance. This of course might include *not* reacting, bearing in mind that consciously not reacting is a genuine strategic attitude.

So, if one swallow does not make a summer, what was hidden behind the positive results? The facts were straightforward:

- Fewer flights through cutbacks to diminish SN's exposure to the increase in the unit cost of operation, while the quality of revenue was not guaranteed in the long term.
- Revenue boosted by the fuel surcharge and some specific initiatives of price increases.
- Same number of or fewer passengers.

The last point was strategic, while the first two were just paramount good management practices. The lack of growth or even some decline in passenger numbers was strategic and needed to be assessed for all its impact.

Why was SN's offer to the market not continuing to recruit new passengers? Was it a question of sales transformation? Was anybody else getting this growth pattern, including competition or substitution of service, the high-speed train for instance? Did it mean that the market was hitting a ceiling, which our in-depth understanding of the business situation had yet not recognized? One immediately sees that this tool is here to keep feeding a never-ending dialogue about consolidation of the company's competitive position. There is nothing brilliant in this, it just requires us to be systematic and not to find any complacent reason for not doing it. This complacency always has a result in terms of market share or point of RoDoC, because one has lost the initiative of action or reaction.

This trend of stagnating or declining passengers was not specific to one route but was across the whole company. The hypothesis was obvious: the contribution of the passengers could only increase thanks to one of the leverage points, the unit pax contribution; broadly speaking, the price side. Therefore, the potential of contribution increases was highly diminished, but simultaneously the pressure of cost increases could become less and less tenable. So, the idea was to lucidly assess and share with each route management team at which price level of a metric tonne of fuel in Rotterdam the pax contribution of the portfolio of flights on a route would vanish. For instance, the 2006 forecast on the European network with worsening operating conditions anticipated operational performance at 105, with a fuel metric tonne at $656. But an increase of around $70 erased the whole operational contribution. This was just a question of scenario. It was a dreadful one, but according to what we were observing it was not completely unrealistic.

> Deviation analysis guarantees that the unpleasant questions won't be skipped.

At the same period I had a discussion with Sheikh Abdullah Zayed in Dubai, who had many interests in the airline sector. In the cool atmosphere of Le Méridien, a French restaurant, he told me that for 2006 he had among his budget scenarios one with a barrel of oil at $100. The peak at that time was around $65. Meanwhile some relief occurred, but in July 2006 *BusinessWeek* published an article called 'Would $100 Oil Slam the Global Economy'; and as I reread the final manuscript of this book a barrel was peaking at around $75. Therefore, the consequence was simple: no more contribution from the operation and no more money to cover the overheads. But this was not just a working hypothesis. Expensive oil has a good chance of being with us for a long time.

The silver lining in this scenario was that if SN had no more contribution, this handicap was also shared by many of its colleagues. But the silver lining was poor compensation. Fuel surcharges could offset part of the increase, but little by little the whole margin would evaporate and consequently the company needed to anticipate the day when it would be confronted with the reality of its destiny. A forecasting mindset is not an insurance of success, but it guarantees that an opportunity to actively handle the future of the business is not disregarded. In my opinion, this constitutes a significant difference between those who struggle to survive and those who are mortgaging their possibilities of looking at different options. These options are never the easiest, but they need to be read against the ultimate goal of keeping the business afloat.

Managerially speaking, this situation is simple. It forces us to raise unpleasant questions. The nature of current results is usually a sum of plus or minus leading to a positive balance. But when the balance needs to be squared with our nearly-in-real-time performance assessment system, it means that the great contributors of the past contribute less due to the macro-economic pressures of the business, and that simultaneously the not-so-well-performing business units are not able to improve their performance. The reasons are very simple. A poorly performing unit in normal conditions of business cannot enjoy better performance when the business conditions are worsening. Consequently, through a form of management complacency in not recognizing that some business units were absolutely unable to generate an operational contribution – due to the competitive context characterized by overcapacity, worsened by the rising cost of fuel – the whole system was again under severe pressure to do something more radical to protect the future of the business.

The benefit of this suggested forecasting approach is twofold:

- It keeps the business unit management plugged into questioning every day the relevance of the actions implemented to the performance improvement of the business unit.

- It delivers a message that the current business model is heading towards its limits. This does not mean that this business model has lost its relevance, but that its potential for performance improvement is from now on under question. Therefore the dividends of the forecasting approach give senior management the opportunity to think about execution more strategically and more quickly. So, which plan should be devised to keep adjusting the competitive balance of the sector to the advantage of SN?

MACRO-ECONOMIC PERSPECTIVES DICTATE MICRO-MANAGEMENT REACTIONS

A future with high costs of operation sends a clear message that past business conditions are not going to continue and that some appropriate reactions are required. Demonstrated macro-economic pressure forces one to think differently; if not, one runs the risk of the business quickly being out of control.

Business story 14.2: 'The dollar was a better cost cutter'

The macro-economic situation influences the whole picture of micro-management effectiveness. To illustrate this point I need to draw on some experiences nearly 20 years ago. In the mid-1980s, in addition to my academic functions I was deeply involved in the automotive sector through designing and implementing a management development programme called 'From Productivity to Competitiveness' for automotive manufacturer Peugeot. My client was the manufacturing department VP and was also a member of the Executive Committee, the top three or four people in the company, in control of 50 000 people

---

with the help of nearly 200 senior executives. What this team had done was spectacular. Over three to five years, thanks to products that were well appreciated in the market, Peugeot had returned almost from the grave and was back on track.

At the time another company was in virtually the same business situation, Chrysler. We thought it was a good idea to broaden the perspective of the management programme with some insights from the other side of the world. At that time, the company was managed by the widely media-covered Lee Iaccoca, who was assessed as the archetypal example of the turnaround manager for desperate situations. But as in all these spectacular cases of turnaround, behind the scenes there was also a team of outstanding people. One of them was Dick Daught, whom I met in the late 1980s. I got plenty of insight into how this turnaround was run from a manufacturing point of view. A lot of the examples concerned the impact of the mini van, which almost saved Chrysler as the 205 did Peugeot.

Shortly after my return to Europe, I received an issue of *Fortune*. Dick Daught was on the front page with the headline: 'Dick Daught, Iaccoca's Production Whiz'. This long article described the details of Chrysler's manufacturing achievements, which mainly lay in closing outdated facilities, negotiating layoffs and keeping the morale of the troops quite high. But the key point of the article was its conclusion in fewer than 20 words written by *Fortune*'s automotive sector expert Alex Taylor III: 'Dick Daught is a great cost cutter but the dollar was even sharper.' He was referring to the fact that thanks to the evolution of the dollar against the yen, Japanese imported cars had seen their production efficiency-based competitive advantage turn into a handicap. Nevertheless, the next step was already in the pipeline with the wave of Japanese factories being opened in the US in the late 1980s and early 1990s.

At SN, the Chrysler analogy worked in reverse with the fuel price. The macro-economic situation was becoming a structural handicap for the company, which should have forced it to anticipate adjusting its business conditions. Signals of passenger number stagnation in the forecast deviation analysis of each route led us to think differently in relation to costs. It was a temptation to imagine that the solution would emerge from the revenue side. It is difficult not to think in this way first, because it is easier. My proposal with such recurrent warning signals is to diminish the exposure of the company to its most vital concern. By analogy, the new company's tempo becomes a question of minimum performance level. For instance, if one runs 100 metres in above 11 seconds, one is now not even invited to compete in the world championship; 10 years ago the level was 12 seconds. SN faced a similar situation of a bar imposed by the macro-economic context with an impact at the elementary level of strategy execution. This had to be recognized immediately if one did not want to wreck the whole business momentum. Therefore, this dividend of the forecasting tool operates as a comfort breaker, which forces one to look lucidly at the future one is heading towards. One cannot be caught by surprise. What follows illustrutor this principle in the light of the SN case.

The first step consisted in accepting the full consequences of this new business reality by having the courage to write the kind of plan that the vast majority of people would delay until the situation became untenable. Maintaining structurally unprofitable routes or business units is unacceptable. It is a management fallacy to let the positive contribution of the best-performing units be absorbed by the poorly performing ones, with the weak alibi of the consistency of the network or the product protfolio and the value of the customer offer. This management lie was not sustainable.

Forecasting again became a question of dialogue with a very motivating concern: something absolutely had to be done to protect the dividends of the relaunch and to avoid seeing them transformed into a temporary remission. The urgency was to list, discuss and revalidate the hypothesis to agree it for this plan. The hypothesis of forecasted results for 2006 was based on a rather optimistic hypothesis of average oil price at $656 per metric tonne (MT). But the fleet contribution would be completely erased if the MT exceeded $755. In summary, an increase of $20 with the revenue conditions remaining the same would erase 1 point of RoDoC. 2006's forecast was simulated with the following combination of hypotheses:

- 2005's final year-end estimate would certainly head towards nearly zero pax growth at the European network level, with some decreases on a stable perimeter of routes.
- 2005 would increase its revenue thanks to a constant and successful revenue steering effort. The unit pax contribution would keep increasing, an increase, made up of 60 per cent fuel surcharge, the rest a result of the team's yield-management skills. At the flight level, the pax contribution per flight year on year would also increase by one fifth. Up to 21 per cent of this result was also due to capacity management performance (reduction in the number of flights), which allowed the SOF (seat occupancy factor) to be maintained at its 2004 level of 51 pax per flight.

Based on these trends for 2006, there was no market reason to see a dynamic pax growth trend resuming on the existing perimeter of SN. This was just the result of what was observed in the different end-of-year estimation exercises per route and the team's best estimation of the situation at the end of July 2005 for each route. Pax growth has a very strict definition: it means attracting new segments of customers without cannibalizing the existing revenue by seeing, for instance, your existing passengers enjoying lower fares that were initially designed to attract new customers.

If this ideal scheme was not possible, it was therefore highly possible that one would need to run faster to stay in the same place.

This new situation demanded very significant communication from senior management and even the CEO to make it clearly understood to the whole company that the return on effort would be relatively modest. This was what was usually observed when the seat load factor increased due to a very aggressive offer. At the end of the day more passengers had been transported, but with an absence of a real net impact on the bottom line. The strategic intent behind such an approach supposes that passengers who have travelled with one carrier cannot travel with the competition on the same flight. Therefore if this strategy is successful, it ends up controlling the competitor's cash flow and profitability, which in the end fails. This kind of strategic context includes two constraints:

- The difficulty of planning the competitor's exit.
- The limited number of cases where this form of aggressive investment can be made, because the company does not enjoy the resources to do it systematically without falling into bad habits.

The Route Revenue Performance Track Record Graph® drew an annoying conclusion. The vast majority of routes were caught in a volume ceiling trap. It offset the absence of volume growth by a revenue quality effort, which consisted in deploying a defensive strategy at route level. In 2005, the revenue growth posted a 9 per cent increase, which was forecast to become only 2.5 per cent in 2006 resulting from an additional and more risky effort on unit pax contribution. In summary, at the end of July 2005, 2006 had a high risk of heading to nearly zero net revenue growth.

The revenue progression was potentially very modest, so the European network's results became overexposed to the behaviour

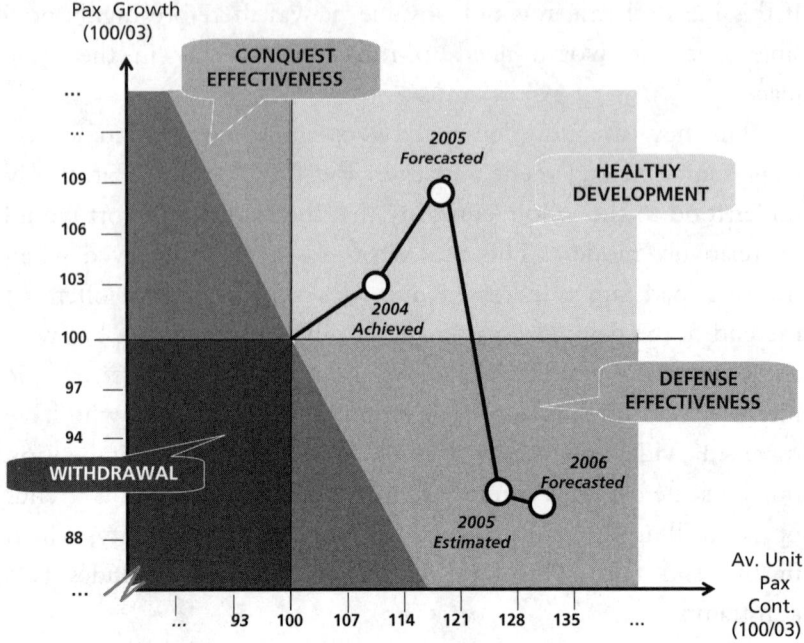

Figure 14.1 Example of a Route Revenue Performance Track Record Graph®.

of costs driven by the oil price rise. The current hypothesis was based on a fleet contribution or RoDoC of 106, which could be erased by an increase in the cost of a metric tonne of oil by $100 to $755. This evolution could also move in the other direction, but that was not the kind of good news that it was wise to expect.

The discussion could also have addressed the challenge from a different angle, observing the consolidated performance of the business units in each strategic group. The Exclusivity group was forecast to generate 100 (index), Match Race 46, Triangular Contest −41 and New routes −26. As in 2004, the 11 Triangular Contest routes and the New routes all lost money, with losses increasing by 19 per cent. The poor performers were even worse, and it was illusory to forecast any substantial route turnaround in these severe business conditions. So faced with such business conditions, what could be done? It was unhealthy to keep operating in such a context without at least having ready a plan for radical change.

This discussion had to be broadened to the consequences for overheads. In this context, if nothing was done, the operating gap (resulting from the difference between the operation contribution and the overheads), which until the end of June 2005 was heading in the correct direction, was forecast to reverse direction and lead to increasing operating losses. The current forecast on the evolution of the contribution from operations legitimately raised the question of the volume of overheads, which had just enjoyed a boost from the relaunch of the company. They had not been managed with the appropriate toughness in order to stick within the benchmark imposed by legacy operators (British Airways, Air France, KLM) or some newcomers such as Ryanair and easyJet. For instance, since the beginning of the YEP report we were assessing the profit zone at a RoDoC of 120 per cent. Early in 2005 profitability was reached at a RoDoC of 123, which meant an increase in overheads of 15 per cent over nearly two years. The operating margin gap was becoming so structural that first it could not be treated with day-to-day management measures, and secondly waiting for natural improvement was illusory. It demanded the immediate consideration of more radical options to stop a vicious circle once and for all.

Consequently, SN had to consider stopping flights to some destinations. In marketing terms a defensive compromise had to be found with some code share partners to protect SN's level of service. But when one addresses this kind of discussion about such radical decisions, there is always a form of company consensus to stick one's head in the sand and say that it is possible but one does not need not to be alarmist. Unfortunately, one always comes back to Andy Grove's principle: only the paranoid survive.

The same phenomenon is observable in politics. For instance, France is a great example where political leaders (until the last presidential election) were telling the country that what you see does not have the meaning you give it if you simply apply educated common sense. This country, if it were not in the European Union, would certainly be like Argentina a few years ago. Its future is plagued by the debt generated by 25 years of utopia that no human life would be enough to pay back, and both right- and left-wing party leaders keep fighting in little fiefdoms, concerned more for their personal power than the interests of the country. It is the same for a company: what counts and should exclusively count is the superior interest of the company, keeping on developing its capacity to survive. It is on this theme that the dialogue must take place, even though it is frightening.

In this specific context the option was very radical: stop flying to some destinations. Based on the current estimated results for 2006, a preliminary list of routes could be generated as applicants for the termination of SN operations. The reason was clear: their turnaround was absolutely unrealistic. By turnaround, one just meant breaking even, a RoDoC of 100, implying that performance stayed 23 points below the operating margin level where the share of overheads was absorbed. Consequently, these routes were assessed as not even reaching an intermediate step created to support a confidence-rebuilding process.

The initial list contained 12 routes; I will just take two examples. In July 2005 I thought it was reasonable to suggest cutting Istanbul and Warsaw. In 2006, year on year the loss per flight had respectively increased by 44 and 100 per cent. So it was futile to hope for any radical improvement for the future unless there was a failure of a competitor, which is always possible but always occurs too late. Moreover, I don't buy the argument of the coherence of the network; this leads nowhere, especially when there is no rational reason to improve. Finally, the management damages are disastrous.

The important lesson in terms of the strategic consequences of the forecasting process was to see that a turnaround plan could become a natural consequence of a forecasting dynamic, and that one needed always to be prepared for this kind of management practice. Noel Goutard, former CEO of the automotive equipment manufacturer Valéo, made this observation when I invited him to speak to the MBA students of EAP when I was dean, 'Yesterday's layoffs are part of today's investments and tomorrow's profits.' If this rule is forgotten it is difficult to stay in the race when the business conditions are worsened by the macroeconomic environment.

BEAR IN MIND

*B*udgets and mid-term strategic plans are common in organizations. Effective forecasting tools are rarer, because they seem redundant and consequently are perceived as less necessary, which is a faulty analysis. The purpose of a forecasting tool is different. It anticipates performance over a broader time perspective, which represents the period over which some substantial competitive evolutions generally occur in one's own industry sector. Forecasting reflects a different business attitude, characteristic of a healthy organization that has already reached a high level of maturity in its relationship to its past performance.

In my different experiences on this topic, I have always developed a forecasting tool for the part of the business that embraces the highest level of uncertainty, revenue. On the revenue side, the team has to demonstrate and continually improve its in-depth insights into the business context, its understanding of the trends and its skills at harvesting and exploiting weak signals. To play an active role in forecasting, managers can't stay stuck in too narrow a definition of their competitive environment – nor in their ivory towers. The forecasting dynamic naturally leads managers out of their offices to sharpen the accuracy of their own skills and fine-tune the correct questions to help them see first their own future performance potential, and consequently their positioning in the industry sector.

A forecasting tool cannot be implemented independently. It logically follows from the development of some other management tools, which lead to an objective corporate dialogue on performance appraisal. A management development sequence is thus necessary, with some prerequisites in order not to fool oneself and not to waste management energy. Therefore, I suggest the following:

- Begin the forecasting process with a systematic end-of-year performance estimate. The goal of doing this is to familiarize the management with the use of performance leverage points.
- Management should validate a ready-to-roll forecasting reference picture, the goal of which is to develop some basic reflexes of deviation analysis, focusing on the same performance leverage points introduced for the end-of-year estimate.
- Monitor corrective measures that should be deployed to stay on track, continually measuring the time between the first signal, the conclusion of the diagnosis and the implementation of the measure.

The value of forecasting does not lie in its accuracy, which is just a consequence, but in the way it forces managers insightfully to bridge the reality of the day-to-day business context and the business situation that is beyond one's own view range. Forecasting is an additional tool and mindset for the CEO in consolidating the individual confidence of management.

Forecasting benefits materialize at the business unit level where, combined with deviation analysis, they help anticipate the correct reaction measures to guarantee staying in line with forecast performance. Forecasting is also a clear acid test at the corporate level. It is a constant reminder to senior executives and the executive committee of what is owed first to one's own micro-management skills and efforts, and secondly to the macro-economic context. Consequently, a further layer of benefits is also observable at the corporate level in the form of a strategic warning system, which can even lead to a restructuring plan to keep protecting the long-term value of the company for its stakeholders.

PRACTICAL REFLEXES FOR SEEING THE FUTURE FIRST

- What is the meaningful timeframe in your industry sector? This should be embraced in order for you to be plugged more relevantly into your industry sector and manage its development tempo more accurately.
- As a CEO, MD or business unit manager, which style of looking at the future do you develop in your company? For instance, how much time do you spend in executive committee meetings sharing evidence of the revenue for each strategic unit? How do you push accountable executives to articulate their commitment to their forecasts?
- Is your own corporate context ready for developing a forecasting approach:
  - Is the elementary level of strategic execution monitoring able to be implemented?
  - Is performance measurement driven by nearly-in-real-time estimates?
  - Do you have a platform for strategic dialogue ready to pull down the walls between the different stakeholders in order to generate performance?
- How much time do you allocate to budget discussions and to forecasting dialogue?
- How often do you discuss budget deviations? How is the time spent during this meeting? What is the timeframe of performance discussed in this session? Is your team focused on the what, the results, or on the how, the way to achieve them?

COMMENTS AND OBSERVATIONS

**Magnus Welander, former CEO of Envirotainer,
currently Business Area President (Europe–Asia)
President of Thulé Car Accessories**

F orecasting is a key sign of absolute organizational health. This is especially true in my own turnaround case, Envirotainer, which in my opinion is radically different from SN's situation of a relaunch. I did not have the comfort of a couple of million euros in the bank, contributed by some new investors. Beyond the early fix-up survival tests, the objective was to be able to deal as fast as possible with the future of the company. That is why I consider forecasting to be a crucial management tool, but also a signal that the company, through the mindset associated with the technique, has turned the page from its past and is therefore plugged into its future.

Numbers that appear to be good form a fatal trap in a turnaround. This can easily happen in budgeting, while forecasting, by its very nature, embraces a broader business perspective.

Forecasting helps the CEO avoid a 'fake' year. In Envirotainer, my goal was to show the genuine results of an effective turn-around. Consequently, I consider forecasting as a form of corporate discipline to stay in line with our own objectives and ambitions – a means to avoid lying to ourselves.

Forecasting is a sign of recovered health because it challenges the timeframe of the business. It anticipates a future that very often is already closer than we think, and that we have not yet recognized because our sensors are not sufficiently developed or trained in its direction. But in this respect, the business model and the sector emphasize the need for this anticipation. For instance in Envirotainer, our business is to rent what we have manufactured or assembled. The availability of these pieces of equipment is what allows us to dominate and keep dominating this niche. But like any service business relying on a costly asset base, the company has to be pretty accurate in terms of asset deployment: too much and we won't take off, not enough and we threaten the future of our service by disappointing customers. Consequently, it is key to push ahead within the time period that is meaningful for your business, always reporting the same 'financially flavoured indicators', which are in fact an assimilated and shared expression of your corporate efficiency.

This forecasting mindset also changes the relationship, more precisely the dialogue, with the board. I must begin with a small nuance: this is in relation to boards that are not dominated by purely finance people. My board has been renewed with CEOs and ex-CEOs. In this context, I can say that forecasting leads to a different dialogue. With forecasting, I indicate to my board how I anticipate the future physiognomy of the business to be. But this is not yet the business. The role of the board is to challenge and to help give that future the highest probability of becoming a business reality. This radically differs from a style of relationship based on 'six months ago you said that . . .' Forecasting is a means, at board level, to work out uncertainty by a genuine future-oriented dialogue. In summary, it helps balance the short-term pressure with what is the best interest of the company.

Forecasting is also a means to save time in an organization. I agree with Jean-Frédéric Mognetti that forecasting cannot happen without systematic deviation analysis. The way the result is achieved is more important than the result itself, and transforming that into exercises for the business unit manager is key. Forecasting is developing a culture of execution with a strategic mindset at the business unit level. But simultaneously, it keeps the senior management plugged into the heart of the everyday business and its behaviour. Consequently, it avoids wasting time in endless strategic reformulation sessions with so little value added, but coming up with nicer words to say the same thing.

When you have been able to develop a strategy, then you have to enforce it – getting the correct one is so important! Forecasting sends you very early on signals of trend breakers that demand an appropriate reaction. But forecasting can also show you an absence of trend breakers and that the business model keeps delivering its performance. There is no need to fix something that isn't broken. This principle is reinforced by the fact that we do believe that the forecasting system is relevant. So its results are reliable. Therefore if there are no signals that the business model is reaching some form of limit, there is no requirement to invent some.

Forecasting gives you spare time when maintaining your business model and becomes a superb organic growth support tool at the senior management level. For instance, it suggests to the executive committee members if the current business stays on track, how will you invest your time to unveil weak signals of additional business growth opportunities?

Finally, when trend breaker signals are occurring the forecasting tool has fulfilled its mission and we are entering another dimension of management. How many weak signals will we need to wait for before we react? Forecasting is a tool that is set up so that we can react without wondering if we have enough information. It avoids the threat of paralysis by analysis. Forecasting helps make 'ready, aim, fire' a reality.

All our business is about value creation, but speed of reaction remains critical. For instance, Envirotainer's business is similar to car rental. We operate via some main stations, which we call global hub airports, where our containers are available and can be picked up and dropped off free of charge. This is the result of traffic management influenced by the presence of our large customers. We also have some stations in far smaller airports, where the customer pays some extra fees to pick up or drop the container, but where the volume is still acceptable to maintain our service. Adapting the status of the station is critical, but how do we anticipate it? Again, we come back to forecasting. If we just see what is happening when the customer traffic has disappeared, we will take the decision to shut the station at this airport, but it will be too late. We need to anticipate, so that we take the decision as early as possible with the correct amount of information. Professor Asi Magnetti explained in Chapter 14, we must have the courage to recognize that our forecasting system is delivering enough information and consequently there is no justification for waiting. This represents in my opion a key characteristic of the capacity of an organizations to build a competitive advantage.

A small variation in the Envirotainer turnaround context must be noted. Empowerment has not been cascaded down to key account managers to make them as accountable as could ideally be expected. So the decision to anticipate the future still ends up on my desk, but we should quickly head towards a more empowered style of management to keep gaining speed of reaction.

CONCLUSION

A turnaround or a relaunch definitely turns the page from the past, but unwittingly or due to insufficiently strong determination this objective can vanish, which gives the past an opportunity to repeat itself. This is absolutely the safest way to fail again!

LEAPING THE MACRO-ECONOMIC HURDLES

What has to be implemented in a turnaround must be perceived in full and has to be radically different in terms of management skills and style. It is a genuine disruption, which must be accepted without question both internally and externally. In SN the senior executive customer survey just nine months after the relaunch reported that this result had already occurred in part of the market. But in this challenge, part of the future success depends on the macro-economic context surrounding the whole operation when the turnaround begins. A turnaround may not be systematically possible, for example, if sufficiently insightful management resources are not available.

Nevertheless, my experience tells me that this execution hurdle is not the most critical. In my opinion, the main constraint to achieving a realistic turnaround is the macro-economic appropriateness of the context to a specific case. I suggest that any turnaround team must be aware of this question; the cost of avoiding it is unsupportable. In *Organic Growth* I advocated the same principle: execution is critical, but one also needs to be in the right place at the right time. This means anticipating how to handle the macro-economic aspect of any business to its benefit. Turnarounds are not exempt from common sense: enjoy the tail wind and develop the talent to perpetuate it. If not . . .

> **The dollar was a better cost cutter than Dick Daught – turnaround demands the passive collaboration of a favourable macro-economic context.**

Macro-economic hurdles take a variety of forms:

- For instance, a lack of key technical resources can easily hamper a project. This is what Caribbean airline BWIA encountered in the lack of availability of maintenance talent. Its own engineers were chased by Trinidad and Tobago's booming gas industry, which pays far higher salaries, on a small island where there is no technical training programme specializing in avionic maintenance and where the airline company has not developed any in-house training because of its endemic crisis.

- In SN's case, it is easy to imagine that if the oil price had reached its current peak (from \$27/barrel in February 2002 to \$75/barrel in July 2006 and over \$98/barrel in October 2007), the project would have run a great risk of staying on the tarmac. The reasons are simple: the conservative revenue fore-cast would not have been sufficient to support the restructuring of all the operations.

- At BWIA, a turnaround platform in its portfolio of routes was clearly identified in 2006. On this basis, a year-on-year comparison at the end of March 2004 gave a RoDoC of 130; in 2006 it was 96, with nearly the same number of passengers and an increased yield (revenue per passenger). The cause was mainly the increase in the oil price, without any hedging protection, and rocketing maintenance costs made worse by ineffective pricing. A recovery option would have been to say that with effective revenue management, the result would have been different. That is true, but it would not have been enough; the improvement would only have been marginal.

In summary, it is fair to bear in mind that when a company is facing a crisis because of some management issues and the problem becomes more complex for macro-economic reasons, the probability of the turnaround succeeding is much lower. The company suffers from a form of corporate septicaemia, and bankruptcy is difficult to avoid. The macro-economic conditions have to be either neutral or favourable to bet on a successful turnaround. In SN's case, the combination of cheap oil prices, cheap asset costs and a unique *raison d'être* – 'Connecting Brussels', the capital of Europe – was unique. Even Swiss did not enjoy the same quality of association of its business fundamentals. So the reborn SN Brussels Airlines benefited from a rather privileged relaunch context after the turmoil of its bankruptcy.

USING THE TOOLKIT

Among the starting conditions of the relaunch of SN, I mentioned the relatively cheap cost of assets. This point is critical. It gives a well-managed company a genuine boost through a relative cost advantage, which allows it to operate in the competitive arena with less risk. But no tree has ever reached the sky! Therefore the next question is: What will happen when this asset has to be renewed? For instance, among the reasons leading to Deutsche Post's purchase of DHL was the renewal of part of its fleet, around 70 aircraft.

> **The assets involved in a turnaround have to be managed with anticipation and insight.**

This question takes us to the tools for micro-managing a turnaround operation, bearing in mind that the ultimate goal is to make a profit, which allows the company to keep its autonomy in renewing its assets. The micro-management toolkit must help us anticipate passing this bar of asset renewal. To fail in achieving it can turn the success enjoyed in the early days of the turnaround into remission, leading in the vast majority of cases to the loss of the company's independence. In my opinion, this asset renewal takes the form of a 'do-or-die' struggle that the board and the executive committee must not complacently skip due to the quick good news of recovered financial health heavily subsidized by exceptional results. That is why the whole experience is bathed in the idea of forward thinking. If a loss of independence does happen this should not be a problem – as long as one realistically anticipates it as the best possible outcome. Therefore the company must take the appropriate care to give true meaning to its day-to-day activities.

The tools used in the SN case followed a logical sequence of first consolidating what the company was inheriting, in order secondly to develop a set of specific tools to put the company back on a healthy growth track. These tools are presented within this ad hoc integrated perspective in this book. But Pierre Henry in his comments and observations on Part III suggests treating the proposed toolbox as a catalogue, from which the CEO can pick the most appropriate tool to prevent or correct declining performance. I urge every manager to follow this advice in order to find some clues to correct any deviation from the roadmap immediately, so as to be back on track without any delay. This discipline is paramount, because when one needs to use this set of integrated tools it is too late for the current management, which has to be substituted by another team. 'The team who leads a company into crisis never tows it out.' This quote from the Chairman of Continental Airlines is so true that CEOs, MDs and even business unit managers should store it among their favourites in their PDAs.

This toolbox has the same characteristics as that developed in *Organic Growth*. The tools suggested are familiar. This does not always mean they have been assimilated, but they are at least understood. Combining them with a new management style prevents them from being rejected and opens up new areas of implementation, which raises confidence that one is back on the correct track. That was why I reintroduced into SN such an obvious tool as BEP analysis, but with a very different flavour when I applied it to a flight instead of a seat. This represented the disruptive side of the use of a familiar tool. The same objective was pursued with the business model concept: I summarized which *modus operandi* the company had selected to make money in its business. This is a simple proposal, but with a heavy burden of accountability. Is it clear for the whole management team? Is it shared with the business doers? Is it hammered into the whole company again and again at any opportunity? A business model must generate money and that is its only true acid test; the rest are merely dreams, which sooner rather than later must be read in the light of the above principle. Finally, a business model aims at generating money, but this is just the consequence of scrupulously coping with its associated drivers. Pierre Henry mentioned his OTP (on time performance) drivers on Sodexho Pass's business model, and added that

> **The basic tools are repellents for fake management practices.**

if they were not achieved, the rest is egocentric marketing or a waste of resources. In the light of this observation SN's turnaround demonstrates that it was a very consistent dynamic.

As of summer 2002, an independent body ranked SN the second most punctual airline in Europe. This had a meaning on which a lot could be leveraged, especially how to coax the limited energy of all the company's talent to maintain such performance. It also indicates that good performance on this business driver was already effective, although the financial results had not yet followed. This inherent business inertia must not cause us to miss the true diagnosis.

In this respect, I propose the contrasting example of BWIA to hammer in the importance of these criteria and of sticking to them. In BWIA's case, due to an endemic maintenance crisis, the first four months of attempts to improve remained stuck at a poor 65 per cent OTP. Organizing a turnaround becomes far more difficult in this context, because one is struggling with performance with the constant risk of falling into the excuse trap – it's somebody else's fault.

The example of DHL is even more striking. Under the umbrella of DPGN (Deutsche Post Global Mail), all the services of the group's different companies were integrated in relation to the salesforce, but this did not follow for operations; or at least, there was such a delay that at the end of 2005 one senior executive of the company said to me angrily, 'In the UK we were missing 3000 pick-ups a week just for system organizational reasons.' He added, 'We are pissing off 3000 customers a week. You don't need a PhD to assess the sales damage.' That is true, especially when less than ten years ago this man and his team were developing a unique sense of customer service based on keeping their promises, meaning faultless pick-up and delivery. The con-

> **The truth of a business model lies in strictly respecting its drivers.**

sequences are clear: being insufficiently obsessed with sticking to the drivers of a business model without any complacency means less money being generated, but also shining talents leaving the company.

Among the proposed tools in SN's 'catalogue', substituting the lens for recognizing the relevant level of elementary strategic execution represented a radical management breakthrough. It introduced in the company the idea that what is analytically correct is not necessarily relevant from a management point of view, especially when the mission has changed so radically. Having a new map with the correct scale is the first step; the next is in relation to its content. The fleet contribution is a second example of how to give a management meaning to something that the company had already used for many years. Initially, very few managers would have spontaneously said that *fleet contribution* was what remains to pay first the overheads, including their own salaries, then the anticipated cost of asset renewal, finally the cost of capital and some extra incentives. Therefore, focusing management attention on an intermediate performance step that raises the question of whether the quality of execution is sufficient to support the overheads radically changes the sense of responsibility. There is nobody else to blame! It is one's own problem. So sharing information with the correct level of focus both questions and motivates.

It is the responsibility of the CEO and his or her team to find the correct formulation of information so that the efforts of the whole company can be read on one simple corporate pulse. The RoDoC, or in SN's case the corporate pulse, is just an application of this principle. It tells a large audience of managers on a daily basis whether or not the company is on track, at the expected pace, with respect to its roadmap.

> **Turnaround tools must be technically correct and relevant to managers, which is not such a natural combination.**

There is an important lesson to be drawn from this in the development of management tools for performance measurement. Turnaround demands turning the page from the past. Taking a dusty performance measurement tool and giving it a new management relevance that can be shared by a far larger audience is a genuine management innovation and consequently a way to turn the page. First, this example says that one leverages what already exists, which is a clear message of lack of arrogance. Secondly, the management use of this term is radically different. This demonstrates that the past is not worth benchmarking, because it was not thoughtful enough to use the tool in this relevant way for the whole company's benefit.

FORECASTING BRINGS SUPERIOR LUCIDITY

Casting into the future is the symptom of a company that has overcome its problems with assessing current performance. If one skips the step of maturing the company in its relationship to its current performance and jumps directly into forecasting, the approach may be fatally built on inconsistencies, which have a great chance of replicating themselves time and again. So it quickly becomes a waste of time.

In SN's case, the future was embraced with an insightful management perspective. Forecasting is not a derivative of strategic planning, it is a core challenge for management in not routinely replicating practices such as the budget, which are correct but won't be insightful enough to prevent repeating the past or to help the company to adapt quickly enough to the evolution of the environment. 'One can always have fake years,' says Magnus

Welander. The threat of this happening must have been considered by one's management system to anticipate the appropriate reaction. Developing a bird's-eye view over the correct period of time, which embraces what is necessary to allow the competitive balance of the sector to materialize, is a powerful answer. This breaks a well-accustomed comfort zone and forces one to adjust the questioning system of one's own management focus. Forecasting is a key signal of organizational health. But it is possible because the strategic dialogue at the business unit level has turned into a form of reflex, which smoothes the speed and the relevance of the exchanges between performance stakeholders.

> **Forecasting is a genuine team management exercise that demonstrates a superior sense of internal dialogue to address one's own future.**

Solid strategic dialogue leading to thoughtful forecasting is in fact the sign of a company that is back in a sound context of growth management. At this stage, the maintenance side of the *Organic Growth* matrix is again under control, as is the sales development. There remains the question of business development, inventing new products or services for new customers. Performance dialogue has become a well-anchored corporate practice. The next step is a dialogue outside the company to harvest weak signals from the market, the fuel of business development.

Consequently, we have in front of us clear evidence that the organization is back in the grip of those who know how to perpetuate growth. They rely on a systematic dialogue internally as well as externally, transformed into a stealthy competitive advantage thanks to their organizational effectiveness. The philosophy of both *Organic Growth* and *Out of the Ashes* asserts that correctly developed tools turned into systematic practices are the safest way to consolidate or recover prosperity.

A CEO COMMITTED TO
HANDS-ON COMMUNICATION

In addition to management tools, *people* are involved in any turn-around or relaunch. The CEO's role is to reignite their passion and redirect it towards where he has recognized that the company can thrive and beat the average performance of the sector. I use the word 'people' and not stereotypes such as 'the company's most valuable assets', because in a turnaround the challenge is to succeed with the available resources and limited room for manoeuvre.

Secondly, I insist that the CEO introduces a competitive benchmarking perspective, because it forces the company to open itself up to the real world and stop considering that ABC and XYZ are not competitors. I have always been very negatively impressed by LEGO's statement in the mid-1980s that it was not concerned by the explosion in electronic games, because these were not a competitor. We clearly saw the consequences of this narrow reading of the real competition situation. To fulfil their role CEOs provide their troops with the right context within which they can operate at the best of their abilities, preventing them from both finding reasons for self-limiting their ambitions – as José Zurstrassen underlines in his comments on Part II – and burning their limited energy on targets that exceed their current resources. So CEOs have to excel at reconciling objectives that do not naturally coexist. It is in this context that the confidence-building mechanism begins. Peter Davies and José Zurstrassen combine their observations and experiences to develop a pilot-like attitude of always looking forward. They tell us that people deserve respect for what they have endured in a bankruptcy, but compassion does not repair the wounds of the past; action and success do. José pushes this point even further in saying that we need to avoid people 'falling in love with their analysis', because doing so prevents them from dealing with what is crucial: moving ahead without any delay, because motion automatically rescales the size of the challenge.

Good CEOs have a talent for boosting people in order to overcome the inertia – even sometimes the paralysis – that can spread through any organization. Monty's recommendation to his generals on the eve of the battle of El Alamein – *you must lead* – is the tempo. So, the turnaround CEO leads the charge, it is their challenge. No genuine success can be obtained by delegation. It is in this context that the empowerment–accountability formula finds all its meaning. The question is what constitutes the CEO's specific leadership in this context. In SN's case, performance accountability occurred first when the reading of performance was actionable, and secondly when the understanding of their own contribution to this performance-generation process became a natural reflex for the vast majority of managers.

> **'You must lead not push.' General Montgomery, December 1942.**

Leadership is about explaining, repeating, hammering the message in and doing it again. Success is reached when people say: What is required from us is obvious. This is the signal that the inertia has been overcome and that leadership has helped change the nature of business performance. This requires how the business model and its performance measurement drivers work to be made transparent. But this is just the prerequisite for an emerging context. The dialogue on how to understand, assess and boost performance effectively creates the oh-so-coveted context of committed acceptance by the doers of the very effective management formula, empowerment–accountability. The lesson is simple: This formula can only work if the steps for implementation have been followed in the correct sequence. In a turnaround time is critical, but that is not a reason to confuse speed and haste, especially when one is addressing such a fundamental issue as building the foundation for the future of the company.

Leveraging a turnaround through people requires a substantial investment in communication. No doubt about it, that is at the heart of the CEO's role and talent. When Sébastien Bazin estimates that this investment burns up to 70 per cent of the CEO's time, he is about right. I think it may be even more, because the CEO by definition operates through communication, dialogue, even demonstration, which people need to stay at the forefront of competitiveness day by day. But effective CEOs have a trick in this respect: they can be everywhere at once. This is something very simple, but not so easy to achieve – management by example. It represents a characteristic of the CEO's mission that a board cannot take for granted if it does not want to run the risk of the casting mistakes evoked by Sébastien Bazin in his observations. José Zurstrassen adds, once the team has moved ahead, 'I am in the middle of the arena with rolled-up shirtsleeves to support the pace.' But in this situation, the obvious conclusion is to say that the CEO is doing that to guarantee results. This is only partly true. The CEO is with people in the eye of the storm to prepare for the next move, to anticipate.

This is Peter Davies spending three-quarters of his time during any intercontinental flight speaking with the crew, either in the cockpit or in the galley; or, with a copy of the flight manifest in one hand and the RoDoC for the day in the other, explaining to a crew on a night stop in the Barcelona Novotel how performance was to be measured, insisting that the perception of half-full or half-empty aircraft had to be replaced by real facts. It is the wife of a distribution chain president complaining to mine that her husband was unable to drive from Paris to the Alps (650 km of motorway) in less than 12 hours because he had always three to four stops at some shops in his chain. It is Sam Walton stopping one of his trucks on the road to get a lift back to Bentonville and discussing logistics with the driver. It is François Michelin in the same overalls as his engineers listening to a supplier presenting an innovation. These are forms of hands-on communication, which secure the present and create the context for the future. Communication is the CEO's pitch to hammer in the sense of what has to be done. Consequently, it is also a matter of discipline at welcoming any opportunity for contact with any stakeholders to remind them of what has to be done, never assuming that it is obvious. It is up to the CEO to find a way to renew their style and their approach so as not to be boring and to create the critical mass of communication beyond which everything becomes easier and possible.

This management journey is now nearly over. Welcome to Belgium. It is 6.15 a.m., we are 5 minutes ahead of schedule, the outside temperature is 28°. It is dazzling. It was our pleasure to make your flight with SN delightful, we are just 'Passionate About You'.[38] Hope to see you soon again on one of our routes!

---

[38] Aspiration (promise what is the best) of SN Brussels Airlines' staff to its clients invented and coached from 2003 to early 2006 by Philip Saunders, SN Brussels Airlines Commercial EVP.

EPILOGUE: ABORTED TURNAROUND

*D*id SN Brussels Airlines achieve an effective turnaround or merely a spectacular remission?

The answer is not easy. SN Brussels Airlines, the company that was resurrected out of the ashes of SABENA, has in fact vanished into its merger with Virgin Express. This phenomenon started far earlier than the formal market merger of the two companies in March 2007. Once Rob Kuijpers, the soul of the turnaround, had left in September 2005, it was quite unrealistic to imagine the momentum being prolonged beyond his replacement. This does not mean that another turnaround story cannot be written, but it will be very different and I do not have the inside knowledge to assess its detailed management relevance. Nevertheless, based on what is known about the airline sector, some final management comments can act as an epilogue.

Don't break the recovery tempo, even unwittingly

A merger is as difficult as a turnaround. SN Brussels Airlines was just sending encouraging signals of recovery when its board assigned it one of the most difficult management challenges, certainly over-rating its barely recovered management skills. Upsetting the tempo of a recovery, even with good intentions, is always costly, strategically and financially.

Mission impossible is not a management practice

With a new company name, Brussels Airlines, it is easy to fool oneself into believing that one is still in the SN Brussels Airlines story. But Brussels Airlines has the legacy of two companies, both in terms of strengths and weaknesses. Unfortunately, mergers follow Gresham's law that the bad chases out the good. Staying in the airline sector, Air France's CEO knew this principle. Early in 2007, he insightfully disregarded Alitalia's case for benefiting from the regulatory conditions offered by the Italian government. Having struggled for three years at executing the most daring merger in airline history was not a reason to overestimate one's own health at embracing an overstretching challenge, unless the deal conditions radically change.

Enforce a relentless sense of performance improvement

The merger between Air France and KLM has executed its initial plan of synergy based on savings or productivity gains that have never been reached before. The project has led for instance to a plan for both companies of €1.4 billion saving over the next three years. As this example reminds us, in a merger plans for cost savings are not wishful thinking. They are a crucial part of the execution, where both partners consolidate their respective healthy positions to contribute to a common goal. A merger is not a magic wand that can erase endemic weaknesses. It requires a little dose of insight and a great deal of perspiration.

Virgin Express had been losing money for many years and had a business model deprived of any successful momentum. As Fortune reported, 'Virgin Express . . . lost $34 million in 2003 and 2004 combined.'[39] Therefore Virgin Express brought its structural handicaps into the merger with SN Brussels Airlines, whatever strategic ambitions were articulated for the merged company or the capital restructuring achieved by Virgin. Unfortunately for SN, even though some exceptional revenue from the SABENA epoch allowed it to keep showing good financial results for 2006, the strategic reality had already sent some annoying warning signals in 2005. The latter year's poor or negative fleet contribution performers had meanwhile not radically improved their status; there was no reason for them to do so. The company was not minded nor committed enough to improve. This does not mean that nothing was done, but the question is whether it was appropriate. Donald Sull[40] very relevantly qualified this kind of business situation as a case of active inertia. A lot of energy is invested in projects that are not always consistent with each other, but together they do not modify the competitive balance of the sector to one's advantage.

[39] Andy Serwer (2005) 'Do Branson's profits equal his joie de vivre?' *Fortune*, 17 October.
[40] Donald Sull (1999) 'Why Good Companies Go Bad', *Harvard Business Review*, August.

Don't become a theme of market conquest for the competition

The competitive environment for Brussels Airlines worsened, with some participants in the market looking even more eagerly for alternative sources of growth to prolong their own successful growth momentum. The competition certainly also did its homework, assessing that SN + Virgin was not the fighter that SN alone had been. The reason was simple: the new company had gained weight and lost nimbleness. It was not fit enough for the challenges it faced. Predictably, the new company's most profitable routes ended up under competitive pressure from easyJet. Geneva–Brussels was an obvious target with very little investment for the attacker, and Nice–Brussels would certainly follow. But unfortunately, in a performance improvement context that is achieving too little progress, even a limited attack immediately put under pressure some 15 per cent of the company's total European fleet contribution. And given that a problem never comes alone, Ryanair from Brussels South is also offering an alternative option to Malaga, one of Virgin's top three performing routes.

These do not sound like encouraging signals when in addition the company's 'fleet contribution cow', Africa, is suffering. The relaunch's business model is caught between the majors, Air France–KLM, their Trojan horse Kenya Airways, and very active local challengers such as Royal Air Maroc and its subsidiary Air Senegal International, not forgetting the pressure of some exotic guerrilla players like Evabora. All this just signals that the future company's cash flow is under higher competitive constraints. This is merely normal business life except for one nuance, the degree to which the senior management team with the insightful advice of its board has anticipated this more demanding game from a mindset of commitment and talent mobilization.

The above symptoms bring us back to Anne Mulcahy's third point (see page XVII), do everything possible so that the cow does not return to the ditch. Active inertia is not an appropriate cure for such symptoms. Nor is insufficient focus due to the new obligations that Brussels Airlines has imposed on itself, to fight on a competitive front that is too broad for its limited resources and talents.

Don't transform a risk into a fixed cost

Rob Kuijpers was conscious of the inherent risk of the merger. This did not mean that the decision made no sense, but that it had to be executed it with an eye to mitigating the risk. That is why he wanted to keep SN and Virgin Express separate in order, among many other aspects, to protect the benefits of the relaunch and also to secure its successful momentum, which he could leverage as a healthy transformation platform for the company's future development ambitions. This would have not prevented the creation of a shared service centre, nor the implementation of all the potential sources of synergy. His attitude expressed a solid sense of corporate responsibility: retaining the possibility of fighting more effectively on a narrower front, or abandoning or rationalizing a position that is too costly to defend effectively. Unfortunately, in the merger just the opposite was chosen: broadening the front with diluted resources, leading also to the dilution of one's own conviction in the demonstrated business model.

Keep your eye on the ball

The former Executive Chairman would certainly have encountered the same competitive context as described above, but with one big difference or advantage: he would have had an organization that was concentrated, focused on a demonstrated business model, with its eye on the ball, to borrow the words of SN's VP Network. In contrast, the merger combined with Rob Kuijpers' departure created for too long a loose context that led to a loss of direction and thinking again through stereotypes.

For instance, what are the strategic routes for Brussels Airlines? The answer was considered to be European capitals, but this was incorrect: nearly all the European capitals except Berlin, Prague and London are mediocre if not poor performers. Why? Because they represent the backyard of the European majors or their alliances. Nevertheless, this was not a strong enough repellent to prevent the Frankfurt route being opened up because it was seen as strategic. Could SN modify the competitive balance to its own advantage with such a specific plan of resource allocation? No, this was completely unrealistic, unless there were some hidden strategic advantages not accessible through common sense. Therefore, the route joined the list of the top poor performers.

A delayed restructuring remains a restructuring to be executed

The turnaround of SN Brussels Airlines was aborted due to the merger with Virgin Express, but sooner or later, if it wants to keep controlling its destiny, the new company will have to make savings through a deep restructuring. Yesterday's restructurings have been achieved, but with a 'compound competitive interest' that is exponentially proportional to the time wasted waiting. Could this have been avoided? The easy answer would have been to leave the task of renewing the fleet to someone else. So the company could quite naturally have joined the list of applicants to consolidate the portfolio of some European majors, as one can see in Spain, Austria or Italy. But, Air France's CEO sent a clear message about the health of the applicants for this role and Lufthansa's CEO showed the depth of his generosity when he was called on to save Swiss. In this context, common sense suggests that there is no other option than to do it alone. At least that would keep Brussels Airlines' future strategic options fully open.

AFTERWORD

Bernard Charles,
CEO President, Dassault Systèmes

I have never been confronted with a turnaround situation. So, when Jean-Frédéric Mognetti asked me what kind of advice I could offer to companies that have just recovered their corporate health, I thought he had knocked at the wrong door. But *Out of the Ashes* plugs you straight into the day-to-day reality of a corporate health recovery and demolishes my preconceived ideas. In our 25 years of continual effort to become and remain the world leader in our sector, Dassault Systèmes has developed some distinctive organizational practices. I think that two of them can be used as a helpful reminder for just-turned-around companies and provide a safer roadmap into their own future.

I always find that a turnaround stretches the company nearly to breaking point. Over a very short period, a radical transformation frees a company from the legacy of its past, which had led it into a nearly fatal crisis. The turnaround period reinvents the organization's effectiveness, but paradoxically what is achieved is not something exceptional. My observation does not diminish the respect deserved by those involved, but does put it into perspective. In this 'do or die' challenge, a turnaround launches a new corporate era – it doesn't punctuate it. A new, correct tempo now prevails in a business context that in the vast majority of cases has become fiercer. The rejuvenated company is back in the competitive arena, clearly visible on any competitor's radar screens, but if the dynamic of the recovery does not continue, this costly won battle may turn into the sunk costs of a temporary, even though brilliant, remission. This is what Jean-Frédéric Mognetti warns us about in the final pages of *Out of the Ashes*.

Dassault Systèmes' expertise in total product life cycle management connects the company as closely as possible with the reality of its daily life. Consequently, we are obliged to be a very straightforward organization, which naturally considers success as a consequence of how we do things, digging into the roots of our organizational effectiveness. To help perpetuate our past success in a sustainable way – the ambition of any company – we have enforced some in-house management practices. One of them concerns our obsession to create the appropriate internal context for nurturing our ambitions. Every year for the past 25 years, since the second year I joined this 10-person start-up with its prestigious name during my military duty right up to now as CEO, I have systematically announced every December: 'next year is a year of transition.' This means a year where radical changes and exceptional organizational achievements need to be effectively implemented if we have the ambition to keep developing in our sector. This management style of planned disorganization prevents any form of organizational complacency from developing and creates a business attitude where the success one is proudest of is the one about to happen. In addition, I can also say by experience that one forgets too often that it is far less costly to stabilize an organization than to overcome its inertia to kick it into gear.

Therefore perpetuating a momentum for success, like achieving a turnaround, starts by paying scrupulous attention to the organization's internal context. In this respect, I echo Peter Davies' words in Part II. Nevertheless, I want to insist on adding: don't hesitate to change your structure to cultivate the whole organization's nimbleness. Success does not belong to an individual person nor to a team but to the whole organization, enabling it to find its own means of rejuvenation and its capacity to anticipate how it can adapt to a fast-evolving business context. If one does not break and reshuffle the structure often enough, perpetuating success loses its dynamics and turns into dogma. Then, one runs a real risk of merely replicating a successful past whose relevance inexorably vanishes. Breaking structures, redesigning the company every year, is one of Dassault Systèmes' keys to prepare its own dynamics of future success. It is not our exclusive property, but we systematically enforce it collectively as a building block of our own effectiveness and culture. Consequently, I advise newly turned-around companies not to wait too long before developing an equivalent organizational practice.

An appropriate internal context is the first step, but to do what? The answer depends on your own ambitions. José Zurstrassen, a genuine high-tech entrepreneur, doesn't see a substantial skills difference between a turnaround manager and a genuine entrepreneur. To expect to have a future in one's own business sector, one needs to keep intact one's entrepreneurial spirit and attitudes. In *Innovation and Entrepreneurship* Peter Drucker addressed one critical characteristic of entrepreneurship when he wrote: 'innovate, create with one objective, dominate one's own sector.'

My experience leads me to the view that the never–ending quest to dominate one's own sector is the best repellent against losing control of one's destiny. Our company has always been about systematically casting itself into a future that is very often beyond our view range. We will have done it for the third time when this book is published. This represents the most secure means of prolonging the momentum. This can be considered as a practice that meets José Zurstrassen's suggestion: 'don't fall in love with your analysis, move to the solution.' Our approach is: 'Don't be stuck by our own past success . . . respect it, but it is always an intermediate step.' For many years, Dassault Systèmes was a world leader in 3D. When we entered the virtual reality game, some Cassandras considered that the cork was being pushed too far out of the bottle when dealing with the challenge of integrating a million parts, for instance in the case of the Boeing 777. Not only did we make it, but we kept pushing the limits further with the Dreamliner, when we proposed to Boeing that we should manage the total life cycle of the machine from design to end of life. We had climbed the Alps through the virtual reality challenge, but this new one was the Himalayas. What drives us is not an unquenchable passion for conquest but a simple question: what will happen if we don't take up this challenge? The answer is crystal clear: the competition would have caught us up and the costs of this wake-up call would have been at least as high, but with a lower probability of regaining leadership. Consequently, you don't keep running ahead of a pack of followers without having a disruptive challenge simmering. This is always the high toll you have to pay to keep control of your own destiny.

Dominating consists for us in permanently trying to see the future first, as Jean-Frédéric Mognetti expresses it in the last part of his book. This led us to question and redefine the relevance of the perimeter of the business we want to be in. The consequence, multiplying the size of our legitimate accessible market by two, by four, could have paralysed a former niche player like us. But on the contrary, these new competitive horizons have become a corralling theme for all our energy and talents. Dominating for Dassault Systèmes is a key word, but one that has never been a word living by itself. It is always combined with humility, when we consider the new business we want to be in, and lucidity with respect to the competition. The Dassault Systèmes recipe enhances the value delivered to our customers, leaves our past competitions at a standstill with respect to this new game we want to offer to the market, and finally, as the icing on the cake, transforms some of yesterday's potential competitors into allies. Newly recovered companies don't struggle to achieve this costly result and then merely have the ambition to stay in the middle of the pack. That would be inconsistent. They must keep questioning whether it makes sense to dominate their current perimeter within which they are operating.

To sum up, my advice to those who are enjoying recently recovered corporate health: keep breaking your structure, redesigning your organization to nurture the turnaround or entrepreneurial spirit intact, and aim at dominating your sector.

INDEX